Complex Variable Methods in Elasticity

A. H. England

Department of Theoretical Mechanics
University of Nottingham

DOVER PUBLICATIONS, INC.
Mineola, New York

Bibliographical Note

This Dover edition, first published in 2003, is a corrected and augmented republication of the work originally published by Wiley—Interscience, a division of John Wiley & Sons Ltd, London, New York, Sydney, and Toronto, in 1971. Completely new to this Dover edition are the "Summary of Formulae" on page 177 and the "Answers to the Examples" on pages 178–191.

Library of Congress Cataloging-in-Publication Data

England, A. H. (Arthur Henry)
Complex variable methods in elasticity / A.H. England.
p. cm.
Rev. ed. of: London ; New York : Wiley-Interscience, 1971.
Includes bibliographical references and index.
ISBN 0-486-43230-0 (pbk.)
1. Elasticity. 2. Boundary value problems. 3. Functions of complex variables. I. Title.

QA931.E56 2003
531'.382—dc22

2003055477

Manufactured in the United States of America
Dover Publications, Inc., 31 East 2nd Street, Mineola, N.Y. 11501

PREFACE

The formulation of the linear theory of elasticity was largely completed by the middle of the nineteenth century. Since this time the theory has been applied to many problems of importance in engineering and found to yield surprisingly good agreement with physical reality. Perhaps the class of problems which has received most attention has been the plane problems of elasticity in view of their direct practical importance and as approximations to three-dimensional situations. This attention has resulted in a great variety of mathematical methods being applied to their solution. By far the most powerful of these methods is the complex variable approach of Kolosov and Muskhelishvili.

The aims of this text are to give a brief description of this method, illustrating the connexion between the most common boundary value problems of two-dimensional elasticity and certain boundary conditions on functions of a complex variable, and finally to describe the techniques of solution of these problems using complex function theory.

Following a brief chapter describing certain results from complex function theory (which may be omitted until required), the basic equations of the plane strain and generalized plane stress problems of elasticity are derived and their solution expressed in terms of functions of a complex variable. The third and fourth chapters illustrate the use of this representation in the solution of elastic boundary value problems for half-planes and circular regions respectively. These chapters may be read independently but the ideas of continuation are developed in Chapter 3 for the algebraically simpler case of the half-plane. The fifth chapter describes the extension of the methods of Chapter 4 to regions which may be conformally mapped onto a circle by rational functions. Each chapter is followed by a number of problems.

Some knowledge of the linear theory of elasticity and an acquaintance with functions of a complex variable would be useful prerequisites for a study of this book.

As it has not been possible to include a comprehensive list of references to the very numerous works in this field those chosen have been selected to illustrate particular points of difficulty or interest arising in the text.

It is a pleasure to record my thanks to my friends at Nottingham, particularly Professor A. J. M. Spencer and Dr W. A. Green, for their interest in this project, to Professor W. Prager of Brown University but for whom this monograph would not have been written, and to Professor E. T. Onat of Yale University for affording me a most stimulating year there during which some of the manuscript was completed. Finally my thanks are due to Mrs E. Burch and Mrs J. C. Lunn for their efficient typing of the manuscript and to my wife for her active encouragement.

Nottingham A. H. ENGLAND

CONTENTS

INTRODUCTION

Functions of a complex variable were introduced into plane elastic problems in 1909 by Kolosov,[33] who, together with Muskhelishvili,[34] inspired a large school of co-workers in the U.S.S.R.* The resulting developments have been described by Muskhelishvili in two outstanding works which are now available in translation.[43, 44] Other excellent accounts of these techniques have been given by Sokolnikoff,[59] Green and Zerna[21] and Milne-Thomson,[41] and further references may be found in these books and in the survey by Teodorescu.[63]

The method of presentation adopted here is as follows. It is shown that the plane problems of linear elasticity reduce to the solution of Navier's displacement equations of equilibrium subject to certain boundary conditions. On writing the Navier equations in complex variable notation a representation is derived for the elastic displacement in terms of two arbitrary holomorphic functions of a complex variable (the complex potentials). The boundary conditions may then be expressed as certain functional equations relating the complex potentials on the boundaries of the elastic body. Since the solution of these functional equations is extremely difficult for bodies of general shape only the simplest regions will be considered. Attention is concentrated primarily on the half-plane and circular regions for which several methods of solution are available.

The most powerful method of solution employs analytic continuation and is illustrated for half-plane problems in Chapter 3 and for circular regions in Chapter 4. Alternatively, in circular regions, the complex potentials may be represented in terms of either power series or Cauchy integrals along the boundary which enable some boundary value problems to be

*This approach was apparently overlooked elsewhere and a subsequent independent use of the method was made by Stevenson,[61] Green[19] and Milne-Thomson.[40]

solved very conveniently. Particular examples are given in Chapter 4. Finally these methods are applied to the class of bodies which may be conformally mapped onto circular regions.

For reasons of space, certain aspects of the plane problems of elasticity have been omitted. In particular, apart from a few references, there is little description of alternative methods of solution or of extensions of the complex variable methods to deal with strips and wedges or general multiply connected regions. Similarly, plane problems for anisotropic bodies are not considered but accounts of this generalization have been given by Lekhnitskii,[35] Green and Zerna[21] and Milne-Thomson.[41] Finally no mention has been made of the equivalence of the plane problems of elasticity to the bending of plates or to the problems of fluid flow nor is a comparison made between this theory and the plane problems of finite elasticity as discussed by Green and Adkins.[20]

1

FUNCTIONS OF A COMPLEX VARIABLE

In this chapter some of the basic definitions and properties of functions of a complex variable are stated as a preliminary to their use in later sections. It is hoped that sufficient detail has been included to enable readers to resolve points of difficulty without frequent recourse to the standard texts on this subject.[7, 28, 43, 44]

1.1 Basic definitions

In the following it will be assumed that all definitions refer to curves and regions lying entirely in the complex plane.

An *arc* is a continuous non-intersecting line which has a continuously varying tangent except at a finite number of points. A *contour* is a simple closed arc, for example an ellipse.

We shall refer to an open connected set in the plane as a *region*. When a region which we denote by S^+ has one or more non-intersecting contours as its boundary, the positive sense of description of each contour is taken to be that for which the region S^+ lies to the left. For example when S^+ is bounded internally by the contours $C_1, C_2, \ldots, C_n$ and externally by the contour C_0, then $C_1, C_2, \ldots, C_n$ have a clockwise sense of description and C_0 anticlockwise. This is illustrated in Figure 1.1 for the case $n = 2$. We

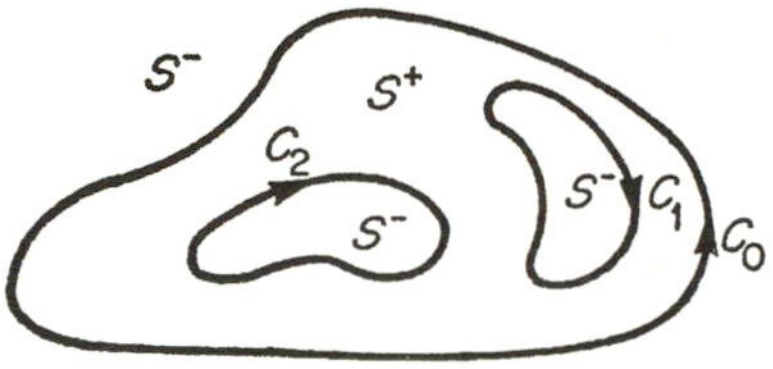

Figure 1.1

denote the open set exterior to S^+ and the bounding contours by S^-, so that on moving in the positive sense along a bounding contour, S^+ lies to the left and S^- to the right. In general S^+ is a multiply connected region, being simply connected only when S^+ is bounded by a single contour C_0.

1.2 Complex functions

Let S be an arbitrary point set in the complex plane, if to each point $z_0 = x_0 + \mathrm{i}y_0$ of S there corresponds a complex number $u(x_0, y_0) + \mathrm{i}v(x_0, y_0)$ we say that a complex function $\theta(z)$ has been defined on S. The value of the function is

$$\theta(z) = u(x, y) + \mathrm{i}v(x, y) \tag{1.1}$$

at the point $z = x + \mathrm{i}y$ where u, v are real functions of the variables x, y. We note that a specific functional dependence on z rather than say $\bar{z} = x - \mathrm{i}y$ is not assumed by this notation. For example $\theta(z) = \bar{z}$ is a complex function.

In view of the relations

$$\begin{aligned} z &= x + \mathrm{i}y, \qquad & \bar{z} &= x - \mathrm{i}y \\ x &= \tfrac{1}{2}(z + \bar{z}), \qquad & y &= 1/2\mathrm{i}\,(z - \bar{z}) \end{aligned}$$

let us *define* the operators $\partial/\partial z$, $\partial/\partial \bar{z}$ as for an ordinary coordinate transformation by the relations

$$\begin{aligned} \frac{\partial}{\partial \bar{z}} &= \frac{1}{2}\left(\frac{\partial}{\partial x} + \mathrm{i}\frac{\partial}{\partial y}\right) \\ \frac{\partial}{\partial z} &= \frac{1}{2}\left(\frac{\partial}{\partial x} - \mathrm{i}\frac{\partial}{\partial y}\right). \end{aligned} \tag{1.2}$$

Then

$$\begin{aligned} 2\frac{\partial \theta}{\partial \bar{z}} &= \left(\frac{\partial}{\partial x} + \mathrm{i}\frac{\partial}{\partial y}\right)(u + \mathrm{i}v) = u_{,x} - v_{,y} + \mathrm{i}(u_{,y} + v_{,x}) \\ 2\frac{\partial \theta}{\partial z} &= \left(\frac{\partial}{\partial x} - \mathrm{i}\frac{\partial}{\partial y}\right)(u + \mathrm{i}v) = u_{,x} + v_{,y} - \mathrm{i}(u_{,y} - v_{,x}). \end{aligned} \tag{1.3}$$

We see that if $\partial\theta/\partial\bar{z} = 0$ at a given point z_0 then the Cauchy–Riemann equations

$$u_{,x} = v_{,y}, \qquad u_{,y} = -v_{,x} \tag{1.4}$$

are satisfied at z_0. Further, if the first partial derivatives of u and v are continuous at z_0, this is a necessary and sufficient condition for the existence of the complex derivative

$$\frac{\mathrm{d}\theta}{\mathrm{d}z} = \theta'(z) = \lim_{\delta z \to 0} \frac{\theta(z + \delta z) - \theta(z)}{\delta z}$$

of $\theta(z)$ at the point z_0. In this case it is simple to show from (1.3) that

$$\frac{\partial \theta}{\partial z} = \theta'(z).$$

Definition A function $\theta(z)$ is said to be *holomorphic*† in a region S^+ if it is single valued in S^+ and its complex derivative $\theta'(z)$ exists at each point of S^+.

For clarity we shall often state when a function is single valued.

1.3 Properties of holomorphic functions

1. If $\theta(z)$ is holomorphic in S^+, then all derivatives of $\theta(z)$ exist and are holomorphic in S^+.

2. If $\theta(z)$ is an arbitrary holomorphic function defined in S^+ then for certain regions S^+ it is possible to use this function to define an associated complex function which is holomorphic in the region which is the image of S^+ in its boundary. This property is of fundamental importance in the method of solution of boundary value problems by continuation. We illustrate this property by defining the associated complex functions for the cases where S^+ is a half plane and a circular region.

Let us denote the half planes $y > 0$ by S^+ and $y < 0$ by S^-. Suppose $\theta(z)$ is holomorphic for $z \in S^+$ then $\theta(\bar{z})$ is defined for all $z \in S^-$ (since for $z \in S^-$, $\bar{z}$ lies in S^+). We now show that the function $\overline{\theta(\bar{z})}$, which is defined for all $z \in S^-$, is holomorphic in S^- and moreover

$$\frac{\mathrm{d}}{\mathrm{d}z}\overline{(\theta(\bar{z}))} = \overline{\theta'(\bar{z})} \quad (z \in S^-). \tag{1.5}$$

From (1.1) and (1.2)

$$\overline{\theta(\bar{z})} = u(x, -y) - \mathrm{i}v(x, -y) \quad (y < 0)$$

and hence

$$\frac{\partial}{\partial \bar{z}}\overline{(\theta(\bar{z}))} = \tfrac{1}{2}\{u_{,x} - v_{,y} - \mathrm{i}(u_{,y} + v_{,x})\} = 0$$

these partial derivatives being evaluated at the point $(x, -y)$, $y < 0$. Hence the associated complex function $\overline{\theta(\bar{z})}$ is holomorphic and by inspection (1.5) may be confirmed.

A similar procedure is possible when S^+ is the circular region $|z| < a$. In this case the image region S^- is $|z| > a$ and for $z \in S^-$ the image point $a^2/\bar{z}$ lies in S^+ (and vice versa). Now if $\theta(z)$ is holomorphic in S^+ then

† The terms analytic and regular are often used.

$\overline{\theta(a^2/\bar{z})}$ is defined for $z \in S^-$ and may be shown to be holomorphic in S^-. In this case however

$$\frac{\mathrm{d}}{\mathrm{d}z}\left(\overline{\theta\left(\frac{a^2}{\bar{z}}\right)}\right) = -\frac{a^2}{z^2}\overline{\theta'\left(\frac{a^2}{\bar{z}}\right)} \quad (z \in S^-). \tag{1.6}$$

Clearly it is possible to interchange the regions S^+ and S^-. Thus if $\theta(z)$ is holomorphic in S^+ $(|z| > a)$ then the associated function $\overline{\theta(a^2/\bar{z})}$ is holomorphic in S^- $(|z| < a)$ and satisfies (1.6) at all points except $z = 0$ where, it will be seen from (1.6), $\overline{\theta(a^2/\bar{z})}$ has an isolated singularity.

3. *The Continuation Theorem.* Suppose $\theta_1(z)$ and $\theta_2(z)$ are holomorphic functions defined in regions S_1 and S_2. Suppose S_1 and S_2 intersect in a domain S and there exists an infinite sequence of distinct points $\{z_n\}$ in S with at least one limit point in S on which

$$\theta_1(z_n) = \theta_2(z_n) \quad (n = 1, 2, \ldots).$$

Then the function

$$\theta(z)\begin{cases} = \theta_1(z) & (z \in S_1) \\ = \theta_2(z) & (z \in S_2) \end{cases}$$

is holomorphic in the union of S_1 and S_2 and $\theta_2(z)$ is the analytic continuation of $\theta_1(z)$ into S_2, $\theta_1(z)$ the analytic continuation of $\theta_2(z)$ into S_1. It often occurs that S_1 and S_2 intersect in a contour L and that along L

$$\theta_1(z) = \theta_2(z) \quad (z \in L).$$

In this case $\theta(z)$ defined as above is holomorphic in $S_1 + S_2 + L$.

4. A holomorphic function may be represented by a unique uniformly convergent power series of the form $\sum_{n=0}^{\infty} \alpha_n(z - z_0)^n$ in the neighbourhood of any point z_0 in its region of holomorphy.

5. We note that if $\theta(z)$ is holomorphic and single valued in the whole plane including the point at infinity then $\theta(z)$ is a constant. This is Liouville's Theorem.

6. *Laurent's Theorem.* If $\theta(z)$ is holomorphic (and single valued) in the annulus $0 < R_1 < |z - z_0| < R_2 < \infty$ then $\theta(z)$ may be represented by a unique uniformly convergent series (the Laurent Series) $\sum_{n=-\infty}^{\infty} \alpha_n(z - z_0)^n$ in the interior of the annulus.

7. *Cauchy's Theorem.* If $\theta(z)$ is a holomorphic function in the region en-

closed by a contour C and is continuous on C then

$$\int_C \theta(z)\, dz = 0.$$

Note that the region enclosed by a single contour C is simply connected.

1.4 Multiple-valued functions

In this monograph we shall restrict our attention to the multiply connected region S^+ which is bounded internally by the set of contours $C_1, C_2, \ldots, C_n$ and externally by the contour C_0 as shown in Figure 1.1. We now determine a representation for the integral of a function which is holomorphic (and single valued) in S^+. For convenience we denote the holomorphic function by $\theta'(z)$ and its integral by $\theta(z)$. If we choose some fixed point z_0 in S^+ then

$$\theta(z) = \int_L \theta'(z)\, dz$$

where L is some arc lying entirely in S^+ and joining z_0 with the current point z. The possibility now exists that by choosing different arcs L in S^+ different values of $\theta(z)$ result.

Let us suppose first of all that S^+ is simply connected (which corresponds to the absence of the internal boundaries $C_1, \ldots, C_n$, i.e. no holes) then if L_1 and L_2 are different paths joining z_0 and z we find

$$\int_{L_1} \theta'(z)\, dz = \int_{L_2} \theta'(z)\, dz - \int_{L_1 - L_2} \theta'(z)\, dz.$$

However since $\theta'(z)$ is holomorphic and single valued within and on the contour $L_1 - L_2$ Cauchy's Theorem (Section 1.3) implies the latter integral is zero. Consequently $\theta(z)$ is independent of the choice of the arc L and is single valued in any simply connected region.

The general multiply connected region S^+ may be made simply connected by introducing n (non-intersecting) cuts joining each of the internal boundaries $C_1, \ldots, C_n$ to the boundary C_0, see Figure 1.2. We

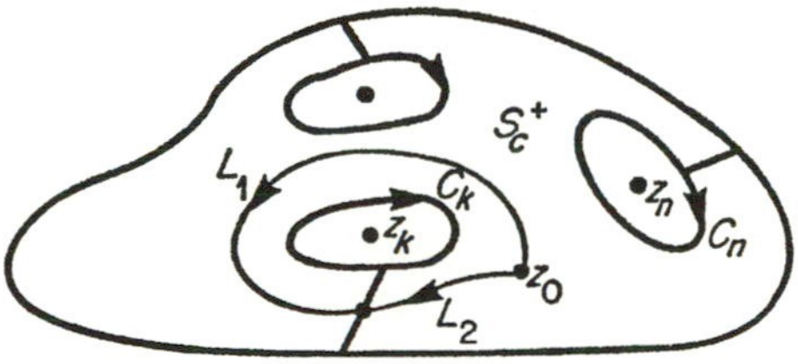

Figure 1.2

denote the cut region by S_C^+. This done it will be seen that $\theta(z)$ is single valued in the cut region S_C^+, but the values of $\theta(z)$ on opposite sides of the cuts will, in general, be different. Consider two arcs L_1 and L_2 lying entirely in S_C^+ which join the fixed point z_0 to corresponding points on opposite sides of the cut between C_k and C_0, see Figure 1.2. Then the change in $\theta(z)$ due to an anticlockwise circuit along the arcs L_1 and L_2 is

$$\int_{L_1 - L_2} \theta'(z)\,\mathrm{d}z.$$

Again we note that $\theta'(z)$ is holomorphic and single valued in S^+ and in particular in the region between the contour $L_2 - L_1$ and C_k, so that by Cauchy's Theorem (Section 1.3),

$$\int_{L_1 - L_2} \theta'(z)\,\mathrm{d}z = -\int_{C_k} \theta'(z)\,\mathrm{d}z = \alpha_k$$

remembering C_k is described clockwise. Thus, for all possible arcs L_1 and L_2 and all points on the cut, $\theta(z)$ increases by a constant α_k in a single anticlockwise circuit of a contour surrounding C_k. Since this type of multi-valuedness holds for each contour C_k $(k = 1, 2, \ldots, n)$, a convenient representation for $\theta(z)$ may be derived.

Consider the function $\log(z - z_k)$ where z_k is a point in the interior of C_k (i.e. outside S^+) then $\log(z - z_k)$ is holomorphic in the cut region S_C^+ and in an anticlockwise circuit around C_k its value increases by $2\pi\mathrm{i}$. Hence the function

$$\theta^*(z) = \theta(z) - \sum_{k=1}^{n} \frac{\alpha_k}{2\pi\mathrm{i}} \log(z - z_k)$$

is continuous across the cuts and so is single valued in S^+. Thus $\theta(z)$ has the representation

$$\theta(z) = \sum_{k=1}^{n} \frac{\alpha_k}{2\pi\mathrm{i}} \log(z - z_k) + \theta^*(z) \tag{1.7}$$

where $\theta^*(z)$ is holomorphic and single valued in the multiply connected region S^+.

Later in the text it will be necessary to integrate partial differential equations of the form

$$\frac{\partial H}{\partial \bar{z}} = 0, \qquad \frac{\partial D}{\partial z} = a(z) \quad (z \in S^+) \tag{1.8}$$

in which $H(z)$ and $D(z)$ are single-valued complex functions in S^+. As some care is required in determining their solutions in the multiply connected region S^+ we examine them in detail here.

Let us write $H(z) = r(x, y) + \mathrm{i}s(x, y)$ and assume the real functions r and s have single-valued continuous first partial derivatives in S^+. From (1.3), $\partial H/\partial \bar{z} = 0$ implies r and s satisfy the Cauchy–Riemann equations (1.4) and hence the complex derivative $H'(z) = r_{,x} - \mathrm{i}r_{,y}$ exists and is single valued in S^+.

If we now assume $H(z)$ is single valued in S^+ we can immediately assert $H(z)$ is holomorphic in S^+. Alternatively, rather than assuming $H(z)$ to be single valued, let us assume its second partial derivatives are continuous and single valued in S^+. In this case it may be confirmed that $H'(z)$ is holomorphic in S^+ and consequently $H(z)$ must be a multiple-valued function in S^+ having a representation of the form (1.7).

The general solution of the homogeneous equation $\partial D/\partial z = 0$ may be derived in a similar manner. On noting that $\partial D/\partial z = 0$ implies $\partial \bar{D}/\partial \bar{z} = 0$ and assuming the first and second partial derivatives of D are continuous and single valued in S^+ we may conclude that $\overline{D(z)}$ is a multiple-valued function of the form (1.7). Thus, under these assumptions, the equation $\partial D/\partial z = 0$ has the solution $D = \overline{\phi(z)}$ where

$$\phi(z) = \sum_{k=1}^{n} \frac{\beta_k}{2\pi \mathrm{i}} \log(z - z_k) + \phi^*(z) \tag{1.9}$$

and $\phi^*(z)$ is holomorphic in S^+.

Let us now consider the more general equation

$$\frac{\partial D}{\partial z} = a(z) \quad (z \in S^+) \tag{1.10}$$

and assume that D and its first and second partial derivatives are continuous and single valued in S^+. To be consistent the complex function $a(z)$ must be assumed single valued in S^+. As $\partial/\partial z$ is a linear operator it will be seen that D is the sum of a particular integral of (1.10), say $D = A(z)$, and a general solution of the homogeneous equation $\partial D/\partial z = 0$ as derived above. Thus the general solution of (1.10) is

$$D = A(z) + \overline{\phi(z)}.$$

Note that as S^+ is multiply connected the possibility exists that both $A(z)$ and $\phi(z)$ are multiple valued, however they must be related so that D is single valued.

Two special cases of (1.10) arise in the text. In the first $\partial D/\partial z = \theta'(z)$, where $\theta'(z)$ is holomorphic in S^+, so that the particular integral is $D = \theta(z)$

(from (1.7)) and the general solution is

$$D = \theta(z) + \overline{\phi(z)}. \tag{1.11}$$

As both $\theta(z)$ and $\phi(z)$ are multiple valued, having representations of the form (1.7) and (1.9), D will be seen to increase by the constant $\alpha_k + \bar{\beta}_k$ in a single anticlockwise circuit of any contour in S^+ surrounding only C_k. Hence as D is required to be single valued the general solution is (1.11) where $\theta(z)$ is given by (1.7) and $\phi(z)$ by (1.9) in which $\beta_k = -\bar{\alpha}_k$ $(k = 1, 2, \ldots, n)$.

In the second case $\partial D/\partial z = \overline{\theta'(z)}$ where $\theta'(z)$ is holomorphic in S^+. Now the particular integral is seen to be $D = z\,\overline{\theta'(z)}$ and is single valued in S^+. Consequently, as D is single valued, the general solution is

$$D = z\,\overline{\theta'(z)} + \overline{\phi(z)} \tag{1.12}$$

where $\phi(z)$ is holomorphic in S^+.

As an illustration of the above points and as a particular example of the theory of Section 2.9 we determine the solution of Laplace's equation

$$\nabla_1^2 u = \left(\frac{\partial^2}{\partial x^2} + \frac{\partial^2}{\partial y^2}\right)u = 0$$

in the region S^+ where u is a real single-valued function of x and y with continuous second partial derivatives. On using the definitions (1.2) we see that

$$\frac{\partial}{\partial \bar{z}}\left(\frac{\partial u}{\partial z}\right) = 0 \quad \text{in } S^+.$$

Hence $\partial u/\partial z$ must be a function which is holomorphic in S^+ which, for convenience, we denote by

$$\frac{\partial u}{\partial z} = \theta'(z).$$

Now from (1.11) the solution of this equation is $u = \theta(z) + \overline{\phi(z)}$ where $\phi(z)$ is an arbitrary function of the form (1.9). Further, as u is real, we must conclude

$$u = \theta(z) + \overline{\theta(z)}.$$

Finally since $\theta(z)$ must have the form (1.7)

$$u = \sum_{k=1}^{n} \frac{1}{2\pi i}\{\alpha_k \log(z - z_k) - \bar{\alpha}_k \log(\bar{z} - \bar{z}_k)\} + \theta^*(z) + \overline{\theta^*(z)}$$

and is single valued in S^+ only if $\alpha_k + \bar{\alpha}_k = 0$ for $k = 1, 2, \ldots, n$. Hence on putting $\alpha_k = \pi i A_k$, where A_k is real, u has the general representation

$$u = \sum_{k=1}^{n} A_k \log|z - z_k| + \theta^*(z) + \overline{\theta^*(z)}$$

where $\theta^*(z)$ is holomorphic in S^+.

In the particular case when S^+ is the annulus $a < |z| < b$, $\theta^*(z)$ may be represented by the Laurent Series $\sum_{n=-\infty}^{\infty} \theta_n z^n$. Consequently u has the general representation

$$u = A \log r + \sum_{n=-\infty}^{\infty} r^n(A_n \cos n\theta + B_n \sin n\theta)$$

within the annulus where $2\theta_n = A_n - iB_n$ and $z = re^{i\theta}$.

1.5 Cauchy integrals

Cauchy integrals on a contour

A. Suppose S^+ is a finite open simply connected region enclosed by the contour C described in an anticlockwise sense. We denote the region exterior to $S^+ + C$ by S^-. Then, if $\theta(z)$ is a complex function holomorphic in S^+ and continuous on C,

$$\frac{1}{2\pi i}\int_C \frac{\theta(t)\,dt}{t - z} = \theta(z) \quad (z \in S^+) \tag{1.13}$$

$$\frac{1}{2\pi i}\int_C \frac{\theta(t)\,dt}{t - z} = 0 \quad (z \in S^-). \tag{1.14}$$

Equation (1.14) is a necessary and sufficient condition that the continuous function $\theta(t)$ defined on C be the boundary value of a function holomorphic in S^+.

B. Suppose $\theta(z)$ is holomorphic in S^- including the point at infinity and continuous on C then

$$\frac{1}{2\pi i}\int_C \frac{\theta(t)\,dt}{t - z} = \theta(\infty) \quad (z \in S^+) \tag{1.15}$$

$$\frac{1}{2\pi i}\int_C \frac{\theta(t)\,dt}{t - z} = \theta(\infty) - \theta(z) \quad (z \in S^-) \tag{1.16}$$

where we again describe C in an anticlockwise sense. The condition (1.15) that the Cauchy integral have a constant value in S^+ is both necessary and sufficient for the continuous function $\theta(t)$ defined on C to be the boundary value of a function holomorphic in S^-.

C. When C is the circle $|z| = a$, necessary and sufficient conditions for the continuous function $\theta(t)$ defined on C to be the boundary value of a function holomorphic in either S^+ or S^- may be written in alternative forms. Suppose $\theta(z)$ is holomorphic in $|z| < a$, then since C is the circle $t\bar{t} = a^2$

$$\frac{1}{2\pi i}\int_C \frac{\overline{\theta(t)}\,dt}{t - z} = \frac{1}{2\pi i}\int_C \frac{\overline{\theta(a^2/\bar{t})}\,dt}{t - z} \quad (z \in S^+)$$

now $\overline{\theta(a^2/\bar{z})}$ is a holomorphic function in $|z| > a$ and hence from (1.15) this integral equals $\overline{\theta(0)}$. Thus

$$\frac{1}{2\pi i}\int_C \frac{\overline{\theta(t)}\,dt}{t - z} = \overline{\theta(0)} \quad (|z| < a) \tag{1.17}$$

a condition which can replace (1.14).

Similarly a necessary and sufficient condition that $\theta(t)$, continuous on the circle $|t| = a$, be the boundary value of a function holomorphic in $|z| > a$ is

$$\frac{1}{2\pi i}\int_C \frac{\overline{\theta(t)}\,dt}{t - z} = 0 \quad \text{for all } |z| < a. \tag{1.18}$$

Hölder conditions

A function $\phi(t)$ defined on an arc L is said to satisfy a Hölder condition on L if for any two points t_1, t_2 on L

$$|\phi(t_1) - \phi(t_2)| < A|t_1 - t_2|^\mu \tag{1.19}$$

where A and μ are positive constants and $0 < \mu < 1$. Suppose $\Phi(z)$ is a complex function defined in the neighbourhood of an arc L. Assume the arc L has a given direction, then about a point t_0 of L (not an end point of L) it is possible to define left and right neighbourhoods of t_0. If $\Phi(z)$ has a unique limit as $z \to t_0$ along any path in the left neighbourhood of t_0 we denote this limit by $\Phi^+(t_0)$. Similarly if a unique limit exists on approach to t_0 from the right neighbourhood it is denoted by $\Phi^-(t_0)$. A function continuous in the neighbourhood of L and for which $\Phi^+(t_0)$, $\Phi^-(t_0)$ exist at all interior points t_0 of L is said to be *sectionally continuous* in the neighbourhood of L.

A function $\Phi(z)$ is said to be *sectionally holomorphic* in a region R cut along the union L of a finite number of non-intersecting arcs if

1. it is holomorphic for all z in $R - L$
2. it is sectionally continuous in the neighbourhood of L
3. it is such that at an end z_0 of an arc in L

$$\lim_{z \to z_0} (z - z_0)\Phi(z) = 0.$$

If a sectionally holomorphic function has a singularity in the form of a branch point at an end z_0 of an arc in L, then since condition 3 is satisfied at z_0 we shall refer to this type of singularity as a *weak singularity*. This type of function plays a fundamental role in two-dimensional elasticity.

Cauchy integral on an arc

Let $\phi(t)$ be a complex function which satisfies the Hölder conditions on an arc L. Then the Cauchy Integral

$$\Phi(z) = \frac{1}{2\pi i} \int_L \frac{\phi(t)\,dt}{t - z} \tag{1.20}$$

may be shown to be a sectionally holomorphic function in the whole plane cut along the arc L. Further the limiting values $\Phi^+(t)$, $\Phi^-(t)$ may be shown to exist on L and satisfy the relations

$$\Phi^+(t_0) - \Phi^-(t_0) = \phi(t_0)$$

$$\Phi^+(t_0) + \Phi^-(t_0) = \frac{1}{\pi i} \int_L \frac{\phi(t)\,dt}{t - t_0} \tag{1.21}$$

where t_0 is a point of L and the integral in (1.21) is interpreted as a principal value. The assumption that $\phi(t)$ satisfies the Hölder condition defined above is sufficient for the existence of the principal value.

These results were first derived by J. Plemelj (1908) and will be referred to as the *Plemelj formulae*. They are derived in Muskhelishvili's *Singular Integral Equations.*[44] We note the Plemelj formulae hold true if we replace the single arc L by a union of several non-intersecting arcs or contours. It should be noted that if L extends to infinity the Hölder conditions as stated above are not sufficient. For instance, if L is an infinite straight line, we must assume in addition that $\phi(t)$ tends to a finite value $\phi(\infty)$ as $t \to \pm\infty$ in such a way that

$$|\phi(t) - \phi(\infty)| = O(|t|^{-\alpha}) \quad (\alpha > 0, \quad t \to \pm\infty). \tag{1.22}$$

We shall refer to this as the Hölder condition at infinity. In this case it may be shown that (1.21) holds and in addition that

$$\Phi(z) \to \pm\tfrac{1}{2}\phi(\infty) \quad \text{as } |z| \to \infty \text{ in } y > 0 \text{ and } y < 0$$

respectively.

1.6 The Hilbert problem

A fundamental problem occurring in two-dimensional elasticity is to determine a sectionally holomorphic function $\Phi(z)$, defined in the whole plane cut along a set of arcs L, having a finite degree k at infinity and satisfying the boundary condition

$$\Phi^+(t) - \kappa\,\Phi^-(t) = \phi(t) \tag{1.23}$$

for all points t on L, where κ is a constant.

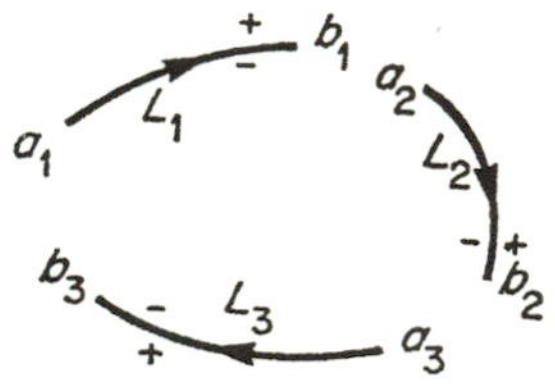

Figure 1.3

We examine first the particular case of determining a sectionally holomorphic function $\Phi(z)$ such that $\Phi(\infty) = 0$ and satisfying the boundary condition

$$\Phi^+(t) - \Phi^-(t) = \phi(t) \quad (t \in L) \tag{1.24}$$

where $\phi(t)$ satisfies the Hölder condition (1.19) on L.

Clearly a solution to this problem is

$$\Phi(z) = \frac{1}{2\pi i}\int_L \frac{\phi(t)\,dt}{t - z} \tag{1.25}$$

since by the Plemelj formulae (1.21) the boundary condition (1.24) is satisfied. This solution is unique since the difference $\Psi(z)$ between two solutions is sectionally holomorphic and satisfies the relation $\Psi^+(t) = \Psi^-(t)$ on L. In particular near each end of L, $\Psi(z)$ has at most a singularity of degree less than 1. Thus since $\Psi(z)$ takes the values $\Psi^+(t) = \Psi^-(t)$ on L it may be shown to be holomorphic everywhere in the finite part of the plane except possibly at the ends of L where there may be isolated singularities of degree less than 1. However such singularities are removable and thus $\Psi(z)$ is holomorphic in the finite part of the plane. Under the assumption $\Phi(z) \to 0$ as $|z| \to \infty$ we must conclude $\Psi(z) = 0$ and hence $\Phi(z)$ is unique.

Clearly if we relax the condition at infinity and look for a solution of finite degree at infinity, $\Psi(z)$ as defined above may be an arbitrary polynomial $P_k(z)$ of degree k. Hence the general solution of (1.24) having a degree k at infinity is

$$\Phi(z) = \frac{1}{2\pi i}\int_L \frac{\phi(t)\,dt}{t - z} + P_k(z). \tag{1.26}$$

In view of this solution it is convenient to refer the boundary conditions (1.24) as *Cauchy boundary conditions*.†

Homogeneous Hilbert problem

Before attempting to solve the general Hilbert problem defined above it is natural to look for solutions to the homogeneous problem. Namely to find a function $X(z)$ sectionally holomorphic in the whole plane cut along L, tending to zero as $|z| \to \infty$ and such that

$$X^+(t) = \kappa X^-(t) \quad (t \in L). \tag{1.27}$$

For simplicity we consider a single arc L with end points a and b oriented from a to b as shown in Figure 1.4. We now show that the branch of the function

$$X(z) = (z - a)^{-\gamma}(z - b)^{\gamma - 1} \tag{1.28}$$

which is such that $\lim_{|z| \to \infty} z\, X(z) = 1$ has these properties.

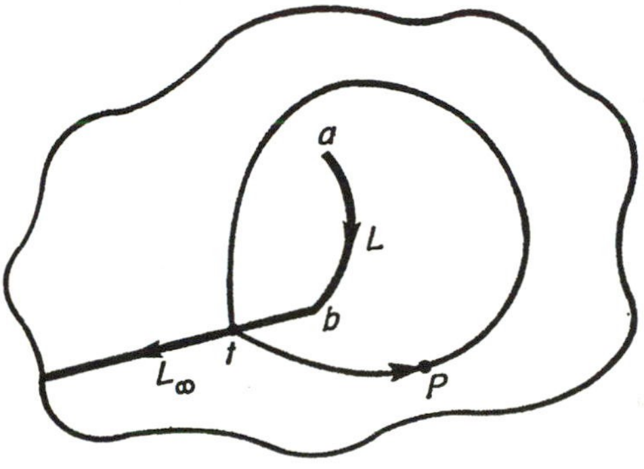

Figure 1.4

As $X(z)$ has branch points at $z = a$, $z = b$ some care must be taken in its definition. We examine first the branch point at $z = b$. If we put $z - b = R_2 e^{i\theta_2}$ then

$$\begin{aligned}(z - b)^{\gamma - 1} &= \exp\{(\gamma - 1)\log(z - b)\} \\ &= e^{(\gamma - 1)\log R_2}\, e^{i(\gamma - 1)\theta_2}\end{aligned}$$

where $e^{(\gamma - 1)\log R_2}$ is uniquely defined but $e^{i(\gamma - 1)\theta_2}$ depends on the value chosen for $\theta_2 = \arg(z - b)$. The argument θ_2 may be chosen uniquely if a cut L_∞ is introduced joining the point b and infinity as in Figure 1.4. The

† There should be no confusion with the Cauchy initial value problem of partial differential equations.

function $(z - b)^{\gamma-1}$ is now uniquely defined and holomorphic in the whole plane cut along L_∞. Similarly for the branch point at $z = a$ if we put $z - a = R_1 e^{i\theta_1}$ then

$$(z - a)^{-\gamma} = e^{-\gamma \log R_1} e^{-i\gamma\theta_1}$$

may be uniquely defined in the whole plane cut along some arc joining a and infinity. To achieve the required properties for $X(z)$ we take this cut to be along $L + L_\infty$.

On using these branches of $(z - a)^{-\gamma}$, $(z - b)^{\gamma-1}$ it is apparent that

$$X(z) = e^{-\gamma \log R_1 + (\gamma-1)\log R_2} e^{-i\gamma\theta_1 + (\gamma-1)\theta_2}$$

is certainly uniquely defined and holomorphic in the plane cut along $L + L_\infty$. We now investigate the change in $X(z)$ as the point z traces a path P beginning at the point t on L_∞ and leading from the left side of L_∞ around the end a to approach the point t from the right side of L_∞ without intersecting $L + L_\infty$. As z traces the path P both θ_1 and θ_2 increase by 2π and hence

$$X^-(t) = X^+(t)e^{-i\gamma 2\pi + i(\gamma-1)2\pi} = X^+(t) \quad (t \in L_\infty).$$

Thus $X(z)$ is continuous across L_∞ and therefore by the continuation theorem $X(z)$ is holomorphic in the whole plane cut along L. It is easily confirmed that this branch of $X(z)$ is such that $z\, X(z) \to 1$ as $|z| \to \infty$.

The same technique may be used to investigate the change in $X(z)$ across the arc L. In this case as we trace a path beginning at a point t on L and leading from the left side of L around the end a to approach t from the right we find θ_1 changes by 2π but θ_2 returns to its original value. Thus

$$X^-(t) = X^+(t)e^{-2\pi i\gamma} \quad (t \in L). \tag{1.29}$$

Hence (1.27) may be satisfied if γ is chosen so that

$$e^{2i\pi\gamma} = \kappa$$

that is

$$2i\pi\gamma = \log|\kappa| + i \arg \kappa. \tag{1.30}$$

Further if the value of arg κ is restricted to lie in the range $0 < \arg\kappa < 2\pi$ then $0 < \text{Re}(\gamma) < 1$ and $X(z)$ has weak singularities at the end points a and b which are such that $\lim_{z\to z_0} (z - z_0)X(z) = 0$ where $z_0 = a$, $z_0 = b$. Thus $X(z)$ is sectionally holomorphic as required and it is convenient to refer to $X(z)$ with this choice of γ as the *basic Plemelj function* for the arc L. The case where κ is real and positive, which has been excluded from the above by the range assumed for arg κ, is examined later.

We now look for the most general solution of the homogeneous Hilbert problem. Suppose

$$F^+(t) = \kappa F^-(t) \quad (t \in L)$$

then we find $F^+(t)/X^+(t) = F^-(t)/X^-(t)$ on L and hence the function $F(z)/X(z) = \Psi(z)$ is holomorphic in the whole plane cut along L across which it satisfies

$$\Psi^+(t) = \Psi^-(t).$$

Thus as above $\Psi(z)$ is holomorphic in the entire plane except possibly at infinity where it can at most have a pole. Hence by Laurent's theorem $\Psi(z)$ is a polynomial and the general solution to the homogeneous Hilbert problem is

$$F(z) = X(z)P(z) \tag{1.31}$$

where $P(z)$ is an arbitrary polynomial. If in addition we require $F(z) \to 0$ as $|z| \to \infty$ then $P(z)$ is a constant.

It should be noticed that, as $X(z)$ has weak singularities at the ends of the arc, $F(z)$ will have such singularities unless $P(z)$ has zeros at one or both of the points a, b. Hence by putting $P(z) = (z - a)Q(z)$, $P(z) = (z - b)Q(z)$ or $P(z) = (z - a)(z - b)Q(z)$ where $Q(z)$ is a polynomial, a set of Plemelj functions can be defined which satisfy the relation (1.27) having either a zero or a weak singularity at a given end. The functions are

$$\begin{aligned} &(z-a)^{-\gamma}(z-b)^{\gamma-1}, \quad (z-a)^{1-\gamma}(z-b)^{\gamma-1}, \\ &(z-a)^{-\gamma}(z-b)^{\gamma}, \qquad (z-a)^{1-\gamma}(z-b)^{\gamma}. \end{aligned} \tag{1.32}$$

These functions have $O(1/z)$, $O(1)$, or $O(z)$ as $|z| \to \infty$, corresponding to two, one, or no singularities.

The product of these functions with an arbitrary polynomial yields the general solution to the homogeneous Hilbert problem with the appropriate behaviour at the ends of the arc. The particular case when κ is real and positive has been excluded so far but may be shown to correspond to

$$F(z) = X_2(z)Q(z)$$

where $X_2(z) = (z - a)^{1-\gamma}(z - b)^{\gamma}$ and $\gamma = \dfrac{1}{2\pi i}\log \kappa$, and of necessity has zeros at the end points a and b.

The Hilbert problem for an arc

The above method indicates a means of solving the more general problem

$$F^+(t) - \kappa F^-(t) = f(t) \quad (t \in L) \tag{1.33}$$

where κ is a constant and $F(z)$ a sectionally holomorphic function in the plane cut along the single arc L with a pole of order k at infinity.

Again dividing through by the basic Plemelj function $X(z)$, from (1.28) and (1.30) the boundary conditions on L become

$$\frac{F^+(t)}{X^+(t)} - \frac{F^-(t)}{X^-(t)} = \frac{f(t)}{X^+(t)}$$

or putting $F(z) = X(z)\Phi(z), \qquad \phi(t) = f(t)/X^+(t)$

$$\Phi^+(t) - \Phi^-(t) = \phi(t) \quad (t \in L).$$

Now using the results (1.26), we find

$$\Phi(z) = \frac{1}{2\pi i}\int_L \frac{\phi(t)\,dt}{t-z} + P(z)$$

where $P(z)$ is an arbitrary polynomial and hence

$$F(z) = \frac{X(z)}{2\pi i}\int_L \frac{f(t)\,dt}{X^+(t)(t-z)} + X(z)P(z). \tag{1.34}$$

It may be shown that if $f(t)$ satisfies the Hölder condition on L this is the general solution to the Hilbert problem. If $F(z)$ has a pole of order k at infinity then $P(z)$ is an arbitrary polynomial of degree $k+1$.

Again it should be noted that $F(z)$ has weak singularities at the end points a and b, however, one or both of these may be removed by use of the appropriate Plemelj function from (1.32) with a resultant increase in the order of $F(z)$ at infinity. Also if κ is real and positive it is necessary to use $X_2(z)$ as the Plemelj function.

The above discussion may be generalized to the case where L is the union of n distinct arcs with ends a_k, b_k for $k = 1, 2, \ldots, n$. In this case the basic Plemelj function corresponding to (1.28) is

$$X(z) = \prod_{k=1}^{n} (z - a_k)^{-\gamma}(z - b_k)^{\gamma-1} \tag{1.35}$$

and we select the branch of $X(z)$ such that $\lim_{|z|\to\infty} z^n X(z) = 1$. Then the solution to the Hilbert problem with boundary conditions (1.33) on the arcs L is (1.34) where $X(z)$ is defined by (1.35) and if $F(z)$ is of degree k at infinity then $P(z)$ is an arbitrary polynomial of degree $k + n$.

Again if zeros are required at the ends c_k, for $k = 1, 2, \ldots, p$, we amend the above solution by using the appropriate Plemelj function in place of $X(z)$ namely, in Muskhelishvili's notation

$$X_p(z) = X(z)(z-c_1)(z-c_2)\ldots(z-c_p). \tag{1.36}$$

Also, as above, if κ is real and positive it is necessary to use $X_{2n}(z)$ in place of $X(z)$ in (1.34) thus removing the singularities from the ends of L.

1.7 Evaluation of the Plemelj functions

In subsequent chapters it will be found necessary to evaluate the Plemelj functions at the origin and infinity. As some care is required in these calculations we derive the results for

$$X(z) = (z-a)^{-\gamma}(z-b)^{\gamma-1} \quad \text{when } \gamma = \tfrac{1}{2} + i\beta \tag{1.37}$$

where $a = r_1 e^{i\alpha_1}$, $b = r_2 e^{i\alpha_2}$ are given points and the branch of $X(z)$ is such that $zX(z) \to 1$ as $|z| \to \infty$.

Let us consider some fixed line through the origin which does not intersect the arc L joining a and b, and suppose z lies at point P on this line, see Figure 1.5. Putting $z - a = R_1 e^{i\theta_1}$, $z - b = R_2 e^{i\theta_2}$ then

$$X(z) = \frac{1}{R_2}\left|\frac{R_2}{R_1}\right|^{\gamma} e^{i\{-\gamma\theta_1 + (\gamma-1)\theta_2\}}.$$

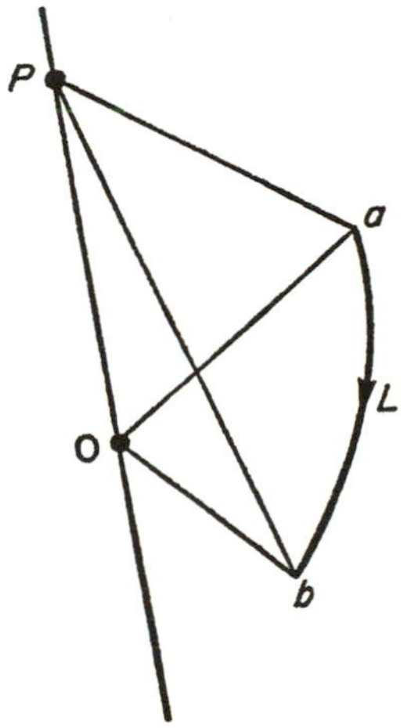

Figure 1.5

The appropriate branch of $X(z)$ has been defined in Section 1.6 and shown to be such that $zX(z) \to 1$ as P tends to infinity. Hence the expansion of $X(z)$ for large $|z|$ is

$$\begin{aligned} X(z) &= \frac{1}{z}\left(1 - \frac{a}{z}\right)^{-\gamma}\left(1 - \frac{b}{z}\right)^{\gamma-1} \\ &= \frac{1}{z}\left\{1 + \frac{\gamma a + (1-\gamma)b}{z} + O\left(\frac{1}{z^2}\right)\right\}. \end{aligned}$$

Alternatively if P moves to the origin along the line we find $R_1 = r_1$, $R_2 = r_2$, $\theta_1 = \alpha_1 + \pi$, $\theta_2 = \alpha_2 + \pi$ so that

$$X(0) = \frac{1}{r_2}\left|\frac{r_2}{r_1}\right|^{\gamma} e^{i\{-\gamma\alpha_1+(\gamma-1)\alpha_2\}-i\pi}$$
$$= -\frac{1}{r_2}\left|\frac{r_2}{r_1}\right|^{\gamma} e^{i\{-\gamma\alpha_1+(\gamma-1)\alpha_2\}}.$$

Thus for small $|z|$

$$X(z) = X(0)\left(1 - \frac{z}{a}\right)^{-\gamma}\left(1 - \frac{z}{b}\right)^{\gamma-1}$$
$$= X(0)\left\{1 + \left(\frac{\gamma}{a} + \frac{(1-\gamma)}{b}\right)z + O(z^2)\right\}. \tag{1.38}$$

In the particular case when $a = ae^{i\phi}$, $b = ae^{-i\phi}$, and we take L to be an arc described clockwise from a to b, we find for large $|z|$

$$X(z) = \frac{1}{z}\left\{1 + (\cos\phi - 2\beta\sin\phi)\frac{a}{z} + O\left(\frac{1}{z^2}\right)\right\} \tag{1.39}$$

and for small $|z|$

$$X(z) = -\frac{1}{a}e^{2\beta\phi}\{1 + (\cos\phi + 2\beta\sin\phi)\frac{z}{a} + O(z^2)\} \tag{1.40}$$

where $\gamma = \frac{1}{2} + i\beta$.

1.8 Evaluation of line integrals

In the solution (1.34) of the Hilbert problem the following line integral occurs

$$I(z) = \int_L \frac{f(t)\,dt}{X^+(t)(t-z)} \tag{1.41}$$

where L is the union of a finite number of arcs $L_1, L_2, \ldots, L_n$ and $X(z)$ is the basic Plemelj function (1.35) satisfying the relation

$$X^+(t) = \kappa X^-(t) \quad \text{on } L. \tag{1.42}$$

If $f(t)$ on L may be expressed as the value on L of a function $f(\zeta)$ which is holomorphic in the neighbourhood of L (say, if $f(t)$ is a polynomial) then the integral along each arc L_k may be expressed in terms of an integral along a lacet C_k surrounding L_k, and the integrals over the lacets evaluated by residue theory, see Figure 1.6.

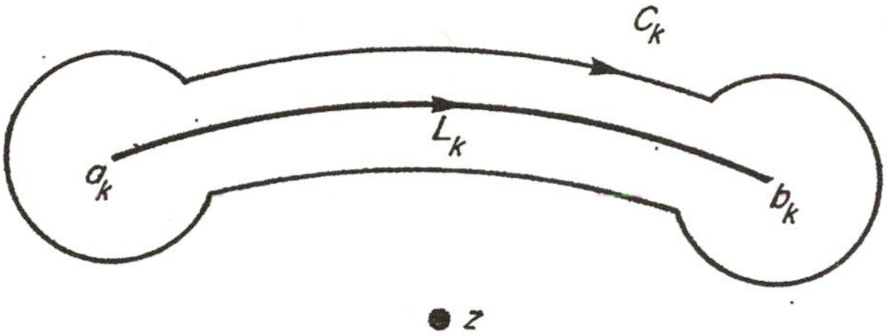

Figure 1.6

Consider the contour integral around the lacet C_k, then

$$\int_{C_k} \frac{f(\zeta)\,d\zeta}{X(\zeta)(\zeta - z)} = \int_{L_k} \frac{f(t)\,dt}{X^+(t)(t - z)} + \lim_{\rho \to 0} \int_{|z - b_k| = \rho} \frac{f(\zeta)\,d\zeta}{X(\zeta)(\zeta - z)}$$
$$- \int_{L_k} \frac{f(t)\,dt}{X^-(t)(t - z)} + \lim_{\rho \to 0} \int_{|z - a_k| = \rho} \frac{f(\zeta)\,d\zeta}{X(\zeta)(\zeta - z)}.$$

Now assuming $f(\zeta)$ is bounded near the ends $\zeta = a_k$, $\zeta = b_k$ it may be shown that the second and fourth integrals above tend to zero as $\rho \to 0$. Hence

$$\int_{C_k} \frac{f(\zeta)\,d\zeta}{X(\zeta)(\zeta - z)} = \int_{L_k} \frac{f(t)}{t - z} \left(\frac{1}{X^+(t)} - \frac{1}{X^-(t)} \right) dt$$

and so from (1.38)

$$= (1 - \kappa) \int_{L_k} \frac{f(t)\,dt}{(t - z)X^+(t)}.$$

The line integral (1.41) may now be expressed in the form

$$I(z) = \int_L \frac{f(t)\,dt}{X^+(t)(t - z)} = \frac{1}{1 - \kappa} \int_C \frac{f(\zeta)\,d\zeta}{X(\zeta)(\zeta - z)}$$

where C is the union of the lacets $C_1, C_2, \ldots, C_n$. This latter integral may be evaluated on replacing C by a circle C_∞, described anticlockwise, at a large distance R from the origin giving

$$I(z) = \frac{1}{1 - \kappa} \left(2\pi i S - \int_{C_\infty} \frac{f(\zeta)\,d\zeta}{X(\zeta)(\zeta - z)} \right)$$

where S is the sum of residues of the poles of $f(\zeta)/X(\zeta)(\zeta - z)$ lying between C and C_∞. If $f(\zeta)$ is a polynomial then

$$I(z) = \frac{1}{1 - \kappa} \left(2\pi i \frac{f(z)}{X(z)} - \lim_{R \to \infty} \int_0^{2\pi} \frac{f(Re^{i\theta})Re^{i\theta} i\,d\theta}{X(Re^{i\theta})(Re^{i\theta} - z)} \right)$$

and it will be seen that only terms independent of $Re^{i\theta}$ in the integrand will contribute to $I(z)$.

As an example suppose $f(t) = 1$ and $X(\zeta) = (\zeta + l)^{-\gamma}(\zeta - l)^{\gamma-1}$ where $2\pi i\gamma = \log \kappa$, then, since the branch of $X(\zeta)$ has been chosen so that $\zeta X(\zeta) \to 1$ as $|\zeta| \to \infty$, $1/X(\zeta)$ has the form

$$\frac{1}{X(\zeta)} = \zeta\left(1 + \frac{l}{\zeta}\right)^{\gamma}\left(1 - \frac{l}{\zeta}\right)^{1-\gamma} = \zeta\left(1 + (2\gamma - 1)\frac{l}{\zeta} + \cdots\right).$$

Hence

$$\begin{aligned} I(z) &= \frac{1}{1-\kappa}\left[2\pi i \frac{1}{X(z)}\right. \\ &\qquad \left. - \lim_{R\to\infty}\int_0^{2\pi}\left(1 - \frac{z}{Re^{i\theta}}\right)^{-1}\{Re^{i\theta} + (2\gamma - 1)l + O(R^{-1})\}i\,d\theta\right] \\ &= \frac{2\pi i}{1-\kappa}\left[\frac{1}{X(z)} - \{z + (2\gamma - 1)l\}\right] \end{aligned} \tag{1.43}$$

2

THE BASIC EQUATIONS OF TWO-DIMENSIONAL ELASTICITY

In this chapter the basic equations of the classical theory of two-dimensional elasticity are formulated. It is shown that the general solution to these equations may be expressed in terms of holomorphic functions of a complex variable. This enables us to apply many of the powerful results of complex function theory discussed in the last chapter to the problems of two-dimensional elasticity. The method owes its foundation to the work of Kolosov[33] and was extended by Muskhelishvili and a group of Russian mathematicians. As mentioned in the introduction an excellent account of this theory has been given by Muskhelishvili.[43] Further accounts have been given by Sokolnikoff,[59] Green and Zerna[21] and Milne-Thomson.[41]

2.1 Basic equations of elasticity†

We consider a homogeneous, isotropic elastic body at a uniform temperature, and suppose that a point of the body originally at a position with coordinates x_i (Latin suffixes will be understood to take the values 1, 2, 3), suffers an infinitesimal displacement u_i with respect to rectangular cartesian coordinate axes. For the present we denote the stress tensor by τ_{ij}, and assume the body contains no stress couples so that τ_{ij} is a symmetric tensor $\tau_{ij} = \tau_{ji}$. The strain ε_{ij} is defined in terms of the displacement gradients by

$$\varepsilon_{ij} = \tfrac{1}{2}(u_{i,j} + u_{j,i}) \tag{2.1}$$

the comma denoting partial differentiation with respect to the appropriate

† The equations quoted in this section may be found in a number of textbooks. An account of the derivation of these equations is given in the first three chapters of Sokolnikoff's book[59] and we have adopted his notation. The relationship between the theories for infinitesimal deformations and finite deformations may be examined in the books of Green and Zerna,[21] Green and Adkins,[20] Truesdell and Toupin,[69] Truesdell and Noll[70].

coordinate, that is, $u_{i,j} = \partial u_i/\partial x_j$. This definition of strain is the usual definition in the infinitesimal theory of elasticity.

The stress–strain relations for the elastic body are taken to be the generalized Hooke's law (see for example Love,[38] Chapter 3)

$$\tau_{ij} = 2\mu\varepsilon_{ij} + \lambda\delta_{ij}\varepsilon_{rr} \tag{2.2}$$

where λ, μ are the Lamé constants, δ_{ij} is the Kronecker delta: $\delta_{ij} = 0$, $i \neq j$; $\delta_{ij} = 1$, $i = j$, and the repeated index summation convention has been adopted, so that

$$\varepsilon_{rr} = \varepsilon_{11} + \varepsilon_{22} + \varepsilon_{33}.$$

Using the definition of strain (2.1) the stress–strain relations take the form

$$\tau_{ij} = \mu(u_{i,j} + u_{j,i}) + \lambda\delta_{ij}u_{r,r}. \tag{2.3}$$

The equations of equilibrium† of the body are

$$\tau_{ij,j} + \rho F_i = 0 \tag{2.4}$$

where F_i is the component of the known body force (for example, gravity or centrifugal force) along the ith direction and ρ is the initial mass density of the body.

If we now substitute the stress–displacement relations (2.3) into the equilibrium equations (2.4), Navier's displacement equations of equilibrium result, namely

$$(\lambda + \mu)u_{r,ri} + \mu u_{i,rr} + \rho F_i = 0. \tag{2.5}$$

We note that a displacement field satisfying (2.5) automatically gives rise, by means of the stress–displacement relations (2.3), to a stress field which satisfies the equations of equilibrium (2.4). Thus the basic equations of linear elasticity are the Navier equations (2.5), which must be solved subject to suitable boundary conditions on the surface of the body, the corresponding stress field being determined by equation (2.3).

In the remainder of this monograph we shall be concerned with two-dimensional elastic problems in which the variable x_3 does not appear in the basic equations or the boundary conditions. These problems are of two distinct types; the first 'plane strain' being concerned with the deformation of a cylindrical body with generators parallel to the x_3-axis under the action of external forces which are such that planes perpendicular to the x_3-axis remain plane. The second type of problem concerns the in-plane extension of a flat thin plate under forces applied around its edges in the

† Alternatively we could consider the equations of motion and group the body force and acceleration terms together, see for example Milne-Thomson,[41] Section 1.1.

plane of the plate. This type of deformation gives rise to the state of 'generalized plane stress'. It will be seen in the following sections that these two classes of problems reduce to the same mathematical formulation and that this may be further simplified on the introduction of functions of a complex variable.

2.2 Plane strain

Using the results of the previous section we shall now derive the basic equations satisfied by a body under the conditions of plane strain. A body is said to be in a state of plane strain when the displacement **u** satisfies the following assumptions at all points of the body

$$\begin{aligned} u_1, u_2 &\text{ are independent of } x_3 \\ u_3 &= 0. \end{aligned} \tag{2.6}$$

Thus plane sections of the body perpendicular to the x_3-axis remain plane. Under these assumptions the stress–displacement relations (2.3) become

$$\tau_{\alpha\beta} = \mu(u_{\alpha,\beta} + u_{\beta,\alpha}) + \lambda\delta_{\alpha\beta}u_{\gamma,\gamma} \tag{2.7}$$

$$\tau_{\alpha 3} = 0$$

$$\tau_{33} = \lambda u_{\gamma,\gamma} \tag{2.8}$$

where we have used the notation that Greek suffixes take the values 1 and 2 and the summation convention has been followed. It will be seen from (2.8) that

$$\tau_{33} = \frac{\lambda}{2(\lambda + \mu)}(\tau_{11} + \tau_{22}) \tag{2.9}$$

and hence, at each point of the body, once $\tau_{11} + \tau_{22}$ has been found, the normal stress τ_{33} sufficient to maintain the body in a state of plane strain is automatically determined.

Applying the assumptions (2.6) to Navier's displacement equations of equilibrium (2.5) we find

$$\begin{aligned} (\lambda + \mu)u_{\gamma,\gamma\alpha} + \mu u_{\alpha,\gamma\gamma} + \rho F_\alpha &= 0 \\ F_3 &= 0 \end{aligned} \tag{2.10}$$

which imply that the body force vector **F** must be independent of x_3 and lie in the plane perpendicular to the x_3 direction. Assuming **F** satisfies these requirements, Navier's equations may be conveniently expressed in a two-dimensional vector notation as

$$(\lambda + \mu)\,\nabla_1(\nabla_1.\mathbf{u}) + \mu\nabla_1^2\mathbf{u} + \rho\mathbf{F} = 0 \tag{2.11}$$

where $\qquad \mathbf{u} = (u_1, u_2), \qquad \mathbf{F} = (F_1, F_2)$

$$\nabla_1 = \left(\frac{\partial}{\partial x_1}, \frac{\partial}{\partial x_2}\right), \qquad \nabla_1^2 = \frac{\partial^2}{\partial x_1^2} + \frac{\partial^2}{\partial x_2^2} \tag{2.12}$$

and we note $\mathbf{u}$ and $\mathbf{F}$ are independent of x_3.

A physical situation corresponding to assumptions (2.6) is the deformation of a uniform cylinder of arbitrary cross-section S^+ with axis and generators parallel to the x_3-axis, under the action of body forces and lateral surface tractions (or displacements) which are independent of x_3 and act parallel to the planes x_3 = constant, see Figure 2.1. As planes normal

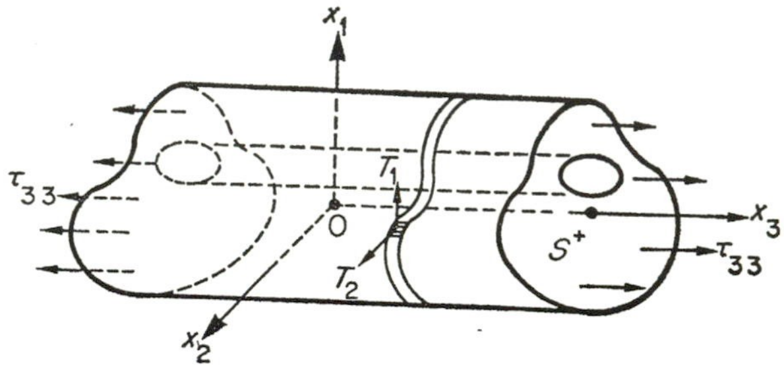

Figure 2.1

to the x_3-axis remain plane, the cylinder must also be constrained by tractions over its end faces in such a way that the longitudinal stress τ_{33} is specified by (2.9). We formulate certain boundary conditions for this deformation in the next section assuming the longitudinal stresses are chosen in this manner to ensure a state of plane strain.

2.3 Boundary conditions

We suppose the cross-section S^+ of the cylinder is bounded by a finite number of contours $C_0, C_1, \ldots, C_n$, with the contour C_0 surrounding all others. The boundary C of the cross-section S^+ is the union of the contours $C_0, C_1, \ldots, C_n$ and we suppose the contours are described so that S^+ always lies to the left. See Figure 1.1. Several physically distinct types of boundary conditions are possible on the surface C of S^+ which are consistent with a plane strain deformation of the cylinder. Of these there are three fundamental types of boundary conditions which seem to be of considerable physical interest.

In the first it is supposed the normal stresses are specified at all points on C and are independent of x_3. Thus if n_1, n_2 are the direction cosines of the exterior unit normal to C, and T_1, T_2 the components along the axes of the external stress, the boundary conditions are

$$\tau_{\alpha\beta} n_\beta = T_\alpha(x_1, x_2) \tag{2.13}$$

for all points (x_1, x_2) in C. We shall refer to the problem of determining a solution of the basic equations (2.7) and (2.11) for plane strain valid throughout the region S^+ and subject to the boundary conditions (2.13) for all points on C as the *stress boundary value problem.*

Alternatively the displacements u_1, u_2 may be specified at all points of C so that the boundary conditions are

$$u_\alpha = g_\alpha(x_1, x_2) \tag{2.14}$$

for all (x_1, x_2) in C. The solution of equations (2.7) and (2.11) subject to boundary conditions (2.14) on C will be referred to as the *displacement boundary value problem.*

In many physical problems the stress boundary conditions (2.13) hold over a part C^* of C and the displacements are defined (2.14) over the remainder $C - C^*$ of C. The solution of the basic equations (2.7) and (2.11) subject to these conditions constitutes the *mixed boundary value problem* and is of a rather more awkward nature than the first two problems formulated above. Clearly more complicated boundary conditions may be considered but these are left until the later sections.

2.4 Generalized plane stress

A second very important class of problems can also be reduced to the equations of plane strain. Let us consider a flat plate $-h \leqslant x_3 \leqslant h$, see Figure 2.2, which is deformed symmetrically about the plane $x_3 = 0$ so that

$$\begin{aligned} &u_\alpha, F_\alpha, \text{ for } \alpha = 1, 2, \text{ are even functions of } x_3 \\ &u_3, F_3, \text{ are odd functions of } x_3. \end{aligned} \tag{2.15}$$

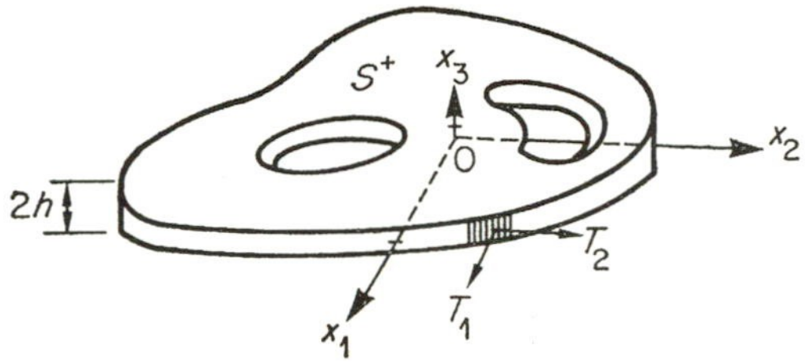

Figure 2.2

Let us now determine the mean stresses and displacements through the thickness of the plate and denote these values by an asterisk so that

$$u_i^* = \frac{1}{2h}\int_{-h}^{h} u_i \, dx_3, \qquad \tau_{ij}^* = \frac{1}{2h}\int_{-h}^{h} \tau_{ij} \, dx_3. \tag{2.16}$$

We see from the assumptions (2.15) that $u_3^* = 0$. If we now integrate the stress–strain relations (2.3) through the thickness of the plate the following relations between the mean stresses and displacements may be obtained on making the assumptions (2.15)

$$\tau_{\alpha\beta}^* = \mu(u_{\alpha,\beta}^* + u_{\beta,\alpha}^*) + \lambda\delta_{\alpha\beta}\Delta^* \tag{2.17}$$

$$\tau_{\alpha 3}^* = \mu(u_{3,\alpha} + u_{\alpha,3})^* = 0$$

$$\tau_{33}^* = \lambda\Delta^* + \frac{\mu}{h}[u_3]_{-h}^{h} \tag{2.18}$$

where

$$\Delta^* = u_{\alpha,\alpha}^* + \frac{1}{2h}[u_3]_{-h}^{h}. \tag{2.19}$$

Further on integrating through Navier's equations (2.5) with respect to x_3 we can find the differential equations satisfied by the mean displacements. These are

$$(\lambda + \mu)\nabla_1(\nabla_1 . \mathbf{u}^*) + \mu\nabla_1^2\mathbf{u}^* + \rho\mathbf{F}^* + (\lambda + \mu)\nabla_1\left(\frac{1}{2h}[u_3]_{-h}^{h}\right) + \frac{\mu}{2h}[\mathbf{u}_{,3}]_{-h}^{h} = 0 \tag{2.20}$$

where the operators ∇_1, ∇_1^2 have been defined in (2.12) and $\mathbf{u}^* = (u_1^*, u_2^*)$, $\mathbf{F} = (F_1^*, F_2^*)$. The third Navier equation being identically zero on integration.

So far no assumptions have been made other than those of symmetry (2.15). Let us now suppose that the shear stresses are zero on the surfaces $x_3 = \pm h$ of the plate and the mean normal stress through the plate is zero so that

$$\tau_{\alpha 3} = 0, \qquad x_3 = \pm h$$
$$\tau_{33}^* = 0. \tag{2.21}$$

It should be noticed that assumptions (2.15) imply τ_{33} is an even function of x_3 so that $\tau_{33}^* = 0$ is a strongly restrictive physical assumption. These relations imply

$$u_{3,\alpha} = -u_{\alpha,3} \qquad \text{on } x_3 = \pm h \tag{2.22}$$

and from (2.18) that

$$\frac{1}{2h}[u_3]_{-h}^{h} = -\frac{\lambda}{\lambda + 2\mu} u_{\alpha,\alpha}^*. \tag{2.23}$$

If we now substitute (2.22), (2.23) into the mean stress–displacement relations (2.17) and the corresponding Navier's equations (2.20) we find

$$\tau^*_{\alpha\beta} = \mu(u^*_{\alpha,\beta} + u^*_{\beta,\alpha}) + \lambda^* \delta_{\alpha\beta} u^*_{\gamma,\gamma} \tag{2.24}$$

$$(\lambda^* + \mu)\nabla_1(\nabla_1 . \mathbf{u}^*) + \mu\nabla_1^2\mathbf{u}^* + \rho\mathbf{F}^* = 0 \tag{2.25}$$

where

$$\lambda^* = \frac{2\lambda\mu}{\lambda + 2\mu}. \tag{2.26}$$

It will be seen that these equations are identical with the corresponding equations for plane strain (2.7) and (2.11) on replacing λ^* by λ.

Thus in a symmetrical deformation of a uniform plate which satisfies conditions (2.15) and (2.21) the equations satisfied by the mean stresses and displacements are identical with those of plane strain.

The idea of introducing the mean stresses and displacements originated with Filon[16] who noted that in the case of a symmetrical deformation of a thin plate, where h is small compared with any other natural length in the plate or the boundary conditions, the variations of the stresses and displacements through the plate would be small and hence the mean values would afford reasonable approximations to actual values in the thin plate. However before we can apply equations (2.24), (2.25) with confidence we must show that the assumption $\tau^*_{33} = 0$ is justified in this case.

Let us suppose that the plate is a cylinder of height $2h$ and that it is thin in the sense defined above. Further let us assume the faces $x_3 = \pm h$ of the plate are free from applied stress and that the surface forces acting on the lateral surfaces of the cylinder and the body force $\mathbf{F}$ are parallel to the plane $x_3 = 0$ and symmetrically distributed with respect to it, so that in particular $F_3 = 0$. We describe this physical situation by saying that the plate is in a state of generalized plane stress and note that conditions (2.15) and the first of (2.21) above are satisfied. Thus

$$\tau_{\alpha 3} = \tau_{33} = 0 \quad \text{on } x_3 = \pm h. \tag{2.27}$$

If we examine the third equation of equilibrium (2.4)

$$\tau_{13,1} + \tau_{23,2} + \tau_{33,3} = 0$$

we see that equations (2.27) imply

$$\tau_{\alpha 3,\alpha} = 0 \quad \text{on } x_3 = \pm h$$

and so $\tau_{33,3} = 0$ on $x_3 = \pm h$. Thus both τ_{33} and its derivative with respect to x_3 vanish on $x_3 = \pm h$. For a thin plate it seems reasonable to assume that τ_{33} differs only slightly from zero throughout the plate and hence the mean value τ_{33}^* is approximately zero. If we assume $\tau_{33}^* = 0$ the mean stresses and displacements satisfy the corresponding Navier's equations (2.25) and the stress–displacement relations (2.24). In many texts it is assumed $\tau_{33} = 0$ throughout the plate which introduces no further restrictions in the basic equations (2.24) and (2.25).

Thus we see the basic equations of the theory of generalized plane stress are of the same form as those of the theory of plane strain. It may be shown similarly that the boundary conditions are equivalent to those formulated in Section 2.3.

Suppose we consider an element of the edge of the plate of width ds and height $2h$. If we integrate with respect to x_3 between $-h$, h the resultant applied stress has components $T_1^* 2h\, ds$, $T_2^* 2h\, ds$ along the x_1 and x_2-axes and has zero out-of-plane bending moments and stress in the x_3 direction under the assumptions stated above. Hence the boundary conditions for the mean stresses are

$$\tau_{\alpha\beta}^* n_\beta = T_\alpha^*$$

where n_1, n_2 are the direction cosines of the exterior unit normal to the edge of the plate. It is clear that displacement and mixed boundary conditions holding in the plane of the plate will also result in boundary conditions for the mean stresses and displacements which are identical to those given in Section 2.3.

In the subsequent sections we shall make no formal distinction between the plane strain and generalized plane stress situations, but deal always with the equations of plane strain, the corresponding results for generalized plane stress being obtained on replacing λ by λ^* and interpreting the stresses and displacements as their mean values.

2.5 Complex variable formulation

It is convenient to repeat at this point the basic equations of two-dimensional elasticity namely Navier's equations

$$(\lambda + \mu)\nabla_1(\nabla_1 . \mathbf{u}) + \mu\nabla_1^2\mathbf{u} + \rho\mathbf{F} = 0 \tag{2.28}$$

and the stress–displacement relations

$$\tau_{\alpha\beta} = \mu(u_{\alpha,\beta} + u_{\beta,\alpha}) + \lambda\delta_{\alpha\beta}u_{\gamma,\gamma} \tag{2.29}$$

these equations holding at all points (x_1, x_2) in the region S^+ occupied by the body. We assume S^+ is a multiply connected region being bounded internally by the contours $C_1, C_2, \ldots, C_n$ and externally by a contour C_0, see Figure 1.1.

Several methods of solution of these (or the equivalent) equations are available. The classical method relying on the construction of the Airy stress function and its subsequent representation in terms of functions of a complex variable has been given in Muskhelishvili's book[43]† and is covered in the examples at the end of the chapter. In this method care has to be taken to ensure the solutions obtained are general solutions for multiply connected regions S^+ and it is hoped this point has been covered in Example 2. A second method of solution based on the Helmholtz decomposition of a two-dimensional vector and the analogue of the method used by Mindlin[42] to derive the Papkovich–Neuber potential functions in three-dimensional elasticity is given in the appendix to this chapter and Example 1.

The method of solution employed here follows directly from expressing the above equations in terms of the complex variable z and stipulating that $\mathbf{u}$ is single valued in S^+. If we put

$$z = x_1 + \mathrm{i}x_2, \qquad \bar{z} = x_1 - \mathrm{i}x_2$$

and define the complex displacement

$$D = u_1 + \mathrm{i}u_2 \tag{2.30}$$

and the complex body force $F = F_1 + \mathrm{i}F_2$, then using the *definitions* of the operators $\dfrac{\partial}{\partial z}, \dfrac{\partial}{\partial \bar{z}}$ given in Section 1.2

$$2\frac{\partial}{\partial z} = \frac{\partial}{\partial x_1} - \mathrm{i}\frac{\partial}{\partial x_2}, \qquad 2\frac{\partial}{\partial \bar{z}} = \frac{\partial}{\partial x_1} + \mathrm{i}\frac{\partial}{\partial x_2} \tag{2.31}$$

Navier's equations (2.28) become

$$2(\lambda + \mu)\frac{\partial}{\partial \bar{z}}\left(\frac{\partial D}{\partial z} + \overline{\frac{\partial D}{\partial z}}\right) + 4\mu\frac{\partial^2 D}{\partial \bar{z}\partial z} + \rho F = 0 \tag{2.32}$$

where the bar denotes the conjugate complex number. For convenience at this point we shall assume it is possible to express the body force F in the form

$$\rho F = \frac{\partial V}{\partial \bar{z}} \tag{2.33}$$

where $V(z, \bar{z})$ is a real-valued function so that $V = \bar{V}$. This question is examined in the next section.

† See also Sokolnikoff,[59] Milne-Thomson,[41] Green and Zerna.[21]

Now from (2.32) and (2.33) we see that Navier's equations take the form

$$\frac{\partial H}{\partial \bar{z}} = 0 \tag{2.34}$$

for all points $z \in S^+$. It has been shown in Section 1.4 that provided H and its first partial derivatives are continuous and single valued in S^+ this equation implies H is a holomorphic function in S^+. Hence if all second derivatives of D are continuous in S^+ we can conclude

$$H = 2(\lambda + \mu)\left(\frac{\partial D}{\partial z} + \overline{\frac{\partial D}{\partial z}}\right) + 4\mu\frac{\partial D}{\partial z} + V = \theta'(z) \tag{2.35}$$

where $\theta'(z)$ is a holomorphic function of the complex variable z in S^+. Since this equation involves only the derivative $\partial D/\partial z$ and its conjugate we may calculate the real and imaginary parts of $\partial D/\partial z$ by taking (2.35) and its conjugate. Hence

$$4(\lambda + 2\mu)\mathrm{Re}\left(\frac{\partial D}{\partial z}\right) = -V + \tfrac{1}{2}\{\theta'(z) + \overline{\theta'(z)}\}$$

$$4\mu\ \mathrm{Im}\left(\frac{\partial D}{\partial z}\right) = \frac{1}{2i}\{\theta'(z) - \overline{\theta'(z)}\}.$$

and

$$8\mu(\lambda + 2\mu)\frac{\partial D}{\partial z} = (\lambda + 3\mu)\theta'(z) - (\lambda + \mu)\overline{\theta'(z)} - 2\mu V. \tag{2.36}$$

Equations of this type have been considered in Section 1.4 from where it may be confirmed that (2.36) has the general solution

$$8\mu(\lambda + 2\mu)D = (\lambda + 3\mu)\theta(z) - (\lambda + \mu)z\,\overline{\theta'(z)} - 2\mu X + \overline{\phi(z)}$$

where $\phi(z)$ is an arbitrary function of the form (1.9) and

$$\frac{\partial X}{\partial z} = V(z, \bar{z}). \tag{2.37}$$

We shall assume $X(z, \bar{z})$ is single valued in S^+.

If we now put

$$\theta(z) = \frac{4(\lambda + 2\mu)}{\lambda + \mu}\Omega(z)$$

$$\phi(z) = -4(\lambda + 2\mu)\omega(z)$$

the general solution of Navier's equations (2.32) for a general region S^+ may be written in the form

$$2\mu D = \kappa\,\Omega(z) - z\,\overline{\Omega'(z)} - \overline{\omega(z)} - \frac{\mu}{2(\lambda + 2\mu)}X \tag{2.38}$$

where $\Omega'(z)$, $\omega'(z)$ are holomorphic functions in region S^+, and

$$\kappa = \frac{\lambda + 3\mu}{\lambda + \mu} = 3 - 4\nu \tag{2.39}$$

for plane strain, and

$$\kappa = \frac{\lambda^* + 3\mu}{\lambda^* + \mu} = \frac{3 - \nu}{1 + \nu}$$

for generalized plane stress where ν is Poisson's ratio. Since for most isotropic elastic materials Poisson's ratio takes values in the range $0 < \nu < \frac{1}{2}$ we see that κ satisfies the inequalities

$$1 < \kappa < 3 \tag{2.40}$$

for both plane strain and generalized plane stress.

As indicated in Section 1.4, $\theta(z)$ and $\phi(z)$ must be functions of the type (1.7) and (1.9) and consequently the complex potentials $\Omega(z)$, $\omega(z)$ are multiple-valued functions in S^+ having representations of the forms

$$\begin{aligned} \Omega(z) &= \sum_{k=1}^{n} \gamma_k \log(z - z_k) + \Omega^*(z) \\ \omega(z) &= \sum_{k=1}^{n} \gamma_k' \log(z - z_k) + \omega^*(z) \end{aligned} \tag{2.41}$$

where $\Omega^*(z)$, $\omega^*(z)$ are arbitrary holomorphic functions in S^+. However the complex potentials are not entirely independent as the displacement D in (2.38) must be single valued in S^+. As in Section 1.4, on describing a contour surrounding each of the holes C_k in S^+ it may be confirmed that D is single valued in S^+ if

$$\kappa\gamma_k + \bar{\gamma}_k' = 0 \quad (k = 1, 2, \ldots, n). \tag{2.42}$$

A physical interpretation of the constants γ_k is given in Section 2.9.

If we now express the stress–displacement relations (2.29) in terms of the complex displacement D we find

$$\begin{aligned} \tau_{11} + \tau_{22} &= 2(\lambda + \mu)u_{\gamma,\gamma} = 2(\lambda + \mu)\left(\frac{\partial D}{\partial z} + \overline{\frac{\partial D}{\partial z}}\right) \\ \tau_{11} - \tau_{22} + 2i\tau_{12} &= 4\mu \frac{\partial D}{\partial \bar{z}}. \end{aligned} \tag{2.43}$$

Using the representation (2.38) derived above, the stresses may be expressed in terms of the complex potentials $\Omega(z)$, $\omega(z)$ by

$$\Theta = \tau_{11} + \tau_{22} = 2\{\Omega'(z) + \overline{\Omega'(z)}\} - \frac{\lambda + \mu}{\lambda + 2\mu} V$$

$$\tau_{11} - \tau_{22} + 2i\tau_{12} = -2\{z\,\overline{\Omega''(z)} + \overline{\omega'(z)}\} - \frac{\mu}{\lambda + 2\mu}\frac{\partial X}{\partial \bar{z}}.$$

It should be noted that the forms of $\Omega'(z)$ and $\omega'(z)$ imply the stresses are single valued in S^+.

The stress combinations $\Theta = \tau_{11} + \tau_{22}$ and $\Phi = \tau_{11} - \tau_{22} + 2i\tau_{12}$ were introduced by Kolosov[33] and it will be found later that Θ is invariant under a rotation of the coordinate system about the x_3-direction. It will be seen that the entire stress and displacement field may be calculated from a knowledge of the complex potentials $\Omega(z)$, $\omega(z)$ and the body force potential $X(z)$.

It is convenient at this point to change our notation to that customarily used in two-dimensional problems. We put $x_1 = x$, $x_2 = y$ and denote the displacements and stresses by

$$u_1 = u, \qquad u_2 = v$$

$$\tau_{11} = \tau_{xx}, \qquad \tau_{12} = \tau_{xy}, \qquad \tau_{22} = \tau_{yy}$$

so that

$$z = x + iy, \qquad D = u + iv.$$

Then, in the absence of a body force, the basic equations of two-dimensional elasticity have the general solution

$$\begin{aligned} 2\mu D &= \kappa\,\Omega(z) - z\,\overline{\Omega'(z)} - \overline{\omega(z)} \\ \Theta = \tau_{xx} + \tau_{yy} &= 2\{\Omega'(z) + \overline{\Omega'(z)}\} \\ \Phi = \tau_{xx} - \tau_{yy} + 2i\tau_{xy} &= -2\{z\,\overline{\Omega''(z)} + \overline{\omega'(z)}\} \end{aligned} \tag{2.44}$$

and

$$\tau_{yy} - i\tau_{xy} = \Omega'(z) + \overline{\Omega'(z)} + z\,\overline{\Omega''(z)} + \overline{\omega'(z)}$$

where $\Omega(z)$, $\omega(z)$ are arbitrary functions of the types (2.41), restricted by (2.42), in the region S^+ occupied by the body.†

2.6 The body force potential

It was assumed in the last section that the body force ρF could be represented in the form $\partial V/\partial \bar{z}$ where $V = \bar{V}$. Although this is a restrictive assumption on F it holds true in the physically interesting cases when the body force is due to gravity or a uniform rotational motion. For gravity $F = C$, a complex constant, and hence $V = \rho C\bar{z} + \rho\bar{C}z$ On integration again we find from (2.37) that

$$X(z, \bar{z}) = \rho Cz\bar{z} + \tfrac{1}{2}\rho\bar{C}z^2. \tag{2.45}$$

† This solution has been derived under the assumption that the single-valued displacement field has continuous second-order partial derivatives in S^+.

For a uniform rotational motion with angular velocity ω about the point z_0

$$F = \omega^2(z - z_0)$$

and hence

$$\begin{aligned} V(z, \bar{z}) &= \omega^2\rho(z - z_0)(\bar{z} - \bar{z}_0) \\ X(z, \bar{z}) &= \tfrac{1}{2}\omega^2\rho(z - z_0)^2(\bar{z} - \bar{z}_0). \end{aligned} \tag{2.46}$$

In the following we shall assume that body force terms are not present in the original equations and this will somewhat simplify the mathematics. However, if required, the effects of such terms may be included by use of the particular integral X in equations (2.38) and (2.43) and the corresponding additional terms in the boundary conditions satisfied by $\Omega(z)$, $\omega(z)$ are simply calculated. It will be seen that these additional terms do not alter the nature of the boundary conditions, but merely the actual values on the boundary, and hence may be omitted without loss of generality.

2.7 Resultant force over an arc

We consider an arc AB in the body S^+ having the direction from A to B as its positive direction and normals **n** pointing to the right when moving from A to B, as shown in Figure 2.3.

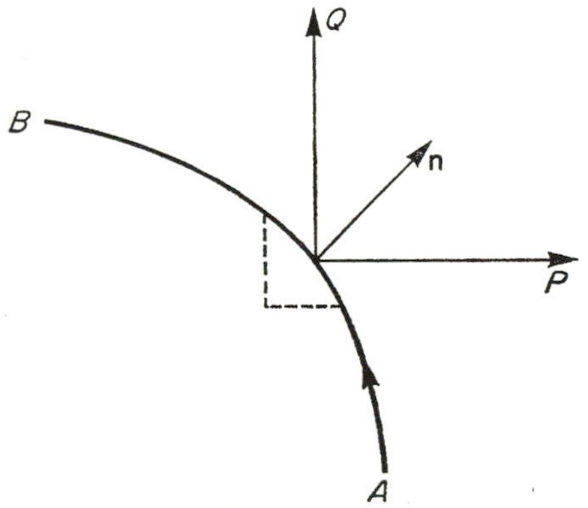

Figure 2.3

We suppose the force which the material on the side of the positive normal exerts on the material to the left of AB over an element of arc ds has components $P\,ds$, $Q\,ds$ along the coordinate axes. Then examining the equilibrium of an elementary triangle we find as in (2.13)

$$P = \tau_{xx}\frac{dy}{ds} - \tau_{xy}\frac{dx}{ds}$$

$$Q = \tau_{xy}\frac{dy}{ds} - \tau_{yy}\frac{dx}{ds}.$$

Hence

$$P + iQ = (\tau_{xx} + i\tau_{xy})\frac{dy}{ds} - i(\tau_{yy} - i\tau_{xy})\frac{dx}{ds}.$$

If we now substitute for the stresses in terms of the complex potentials from (2.44) and group terms we find

$$P + iQ = -i\frac{d}{ds}[\Omega(z) + z\,\overline{\Omega'(z)} + \overline{\omega(z)}].$$

Hence the resultant force over the arc AB acting on material to the left of AB is

$$\int_A^B (P + iQ)\,ds = -i[\Omega(z) + z\,\overline{\Omega'(z)} + \overline{\omega(z)}]_A^B \qquad (2.47)$$

where the square brackets denote the change in the expression as z moves from A to B. This relation clearly extends to the case when the arc AB lies along the boundary C of the material and P and Q are the components of the external force, see Section 2.12.

In a similar manner the resultant moment about the origin of the forces acting over AB may be shown to be

$$M = -\mathrm{Re}\left[z\bar{z}\,\Omega'(z) + z\omega(z) - \int \omega(z)\,dz\right]_A^B. \qquad (2.48)$$

These formulae were first given by Muskhelishvili[43] and are derived in Section 33.

2.8 Arbitrariness of the complex potentials

In this section we investigate how well defined the complex potentials need to be in order to specify a given stress field or a given displacement field.

Suppose that the potentials $\Omega(z)$, $\omega(z)$ give rise to the same stress field as the potentials $\Omega_0(z)$, $\omega_0(z)$. Then the corresponding values of the stress combinations Θ and Φ in (2.44) must be the same at each point of S^+. Hence

$$\mathrm{Re}\{\Omega'(z)\} = \mathrm{Re}\{\Omega_0'(z)\}$$

which, from the Cauchy–Riemann equations, implies

$$\Omega'(z) = \Omega_0'(z) + iA$$

and

$$\Omega(z) = \Omega_0(z) + iAz + \alpha \tag{2.49}$$

where A is a real constant. Similarly the equality of Φ implies

$$\omega(z) = \omega_0(z) + \beta \tag{2.50}$$

where α and β are arbitrary complex numbers.

Corresponding to these two systems the difference in the displacement fields is simply

$$2\mu(D - D_0) = (\kappa + 1)Aiz + \kappa\alpha - \bar{\beta} \tag{2.51}$$

which will be seen to be the combination of a rigid body rotation and a translation.

Since a stress field is determined uniquely from a displacement field, if the two systems $\Omega(z)$, $\omega(z)$ and $\Omega_0(z)$, $\omega_0(z)$ correspond to the same displacement field their differences can be no greater than those indicated in equations (2.49) and (2.50). Further the displacement fields D and D_0 must be identical and hence

$$A = 0, \qquad \kappa\alpha = \bar{\beta}. \tag{2.52}$$

Thus from (2.49) and (2.50) the complex potentials are determined uniquely to within complex constants related by (2.52).

2.9 Restrictions on the complex potentials

In general we have supposed the region S^+ being considered is multiply connected, being bounded internally by contours $C_1, C_2, \ldots, C_n$ and externally by a contour C_0, see Figure 1.1. Under the assumption that the displacement field is single valued† in S^+ we have shown in Section 2.5 that the complex potentials are multiple-valued functions in S^+ of the form

$$\Omega(z) = \sum_{k=1}^{n} \gamma_k \log(z - z_k) + \Omega^*(z) \tag{2.53}$$

$$\omega(z) = \sum_{k=1}^{n} - \kappa\bar{\gamma}_k \log(z - z_k) + \omega^*(z) \tag{2.54}$$

† The assumption of multi-valued displacements leads to the theory of dislocations which has been discussed by Volterra[71] an account of which is given in Love,[38] Section 156*A*. This theory has also been applied to the study of thermal stresses by Muskhelishvili,[43] Chapter 6. See also Nowacki,[47] Chapter 6.

where $\Omega^*(z)$, $\omega^*(z)$ are holomorphic functions in S^+. It has also been confirmed that the stress field is single valued in S^+. On further investigation it turns out that the constants γ_k have a direct physical interpretation in terms of the resultant forces acting over the contours C_k.

If we denote the resultant force applied to the body over the contour C_k by $X_k + \mathrm{i}Y_k$ then from (2.47)

$$X_k + \mathrm{i}Y_k = -\mathrm{i}[\Omega(z) + z\overline{\Omega'(z)} + \overline{\omega(z)}]_{C_k} \tag{2.55}$$

where the contour C_k is described in a clockwise direction. From (2.53) and (2.54) we find

$$X_k + \mathrm{i}Y_k = -2\pi(\kappa + 1)\gamma_k \tag{2.56}$$

and hence the complex potentials take the form

$$\begin{aligned} \Omega(z) &= -\sum_{k=1}^{n} \frac{(X_k + \mathrm{i}Y_k)}{2\pi(1+\kappa)} \log(z - z_k) + \Omega^*(z) \\ \omega(z) &= \sum_{k=1}^{n} \frac{\kappa(X_k - \mathrm{i}Y_k)}{2\pi(1+\kappa)} \log(z - z_k) + \omega^*(z) \end{aligned} \tag{2.57}$$

where $\Omega^*(z)$, $\omega^*(z)$ are single valued in S^+. These are convenient and important formulae.

We note that if the surrounding contour C_0 lies entirely in the finite part of the plane it is simple to show that the resultant force and moment about the origin over C_0 are $-\sum_{k=1}^{n}(X_k + \mathrm{i}Y_k)$ and $-\sum_{k=1}^{n} M_k$ as required by equilibrium where M_k is the resultant moment about the origin over C_k. So far no reference has been made to the case where C_0 extends to infinity though it is apparent that formulae (2.57) hold in this case.

2.10 Conditions at infinity

The infinite plane

We first consider the case where the external contour C_0 has moved to the point at infinity so that the region S^+ is the infinite plane containing holes bounded by the contours C_k $(k = 1, 2, \ldots, n)$. To examine the behaviour of the stress and displacement fields at infinity it is necessary to find expansions of $\Omega(z)$, $\omega(z)$ for large $|z|$. Let us consider the region E defined by $|z| \geqslant R_0$, where the circle $|z| < R_0$ covers all of the contours C_1, $C_2, \ldots, C_n$. Then for $z \in E$, $|z| > |z_k|$ and the logarithmic terms of (2.57) may be written in the form

$$\log(z - z_k) = \log(z) + \log(1 - z_k z^{-1})$$

where the second term is a single-valued holomorphic function in the region E. This may be most easily seen by noting that $\sigma = 1 - z_k z^{-1}$ for $|z| \geqslant R_0$ is defined in a circle of radius $|z_k|/R_0 < 1$ about the point 1. The function $\log(\sigma)$ is holomorphic and single valued in the plane cut along $\arg(\sigma) = \pi$ and hence $\log(1 - z_k z^{-1})$ is single valued for $|z| > R_0$. This is illustrated in Figure 2.4. Since this result holds true for each value

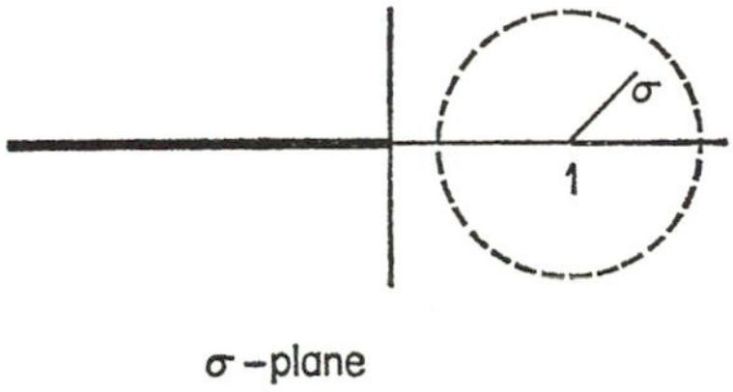

Figure 2.4

of k, if we substitute in (2.57) and sum we find

$$\begin{aligned} \Omega(z) &= -\frac{X + \mathrm{i}Y}{2\pi(1 + \kappa)}\log(z) + \Omega^{**}(z) \\ \omega(z) &= \frac{\kappa(X - \mathrm{i}Y)}{2\pi(1 + \kappa)}\log(z) + \omega^{**}(z) \end{aligned} \tag{2.58}$$

where $\Omega^{**}(z)$, $\omega^{**}(z)$ are single-valued holomorphic functions for $|z| > R_0$ and

$$X + \mathrm{i}Y = \sum_{k=1}^{n} (X_k + \mathrm{i}Y_k) \tag{2.59}$$

is the resultant applied force over all internal boundaries C_k. In view of the properties of $\Omega^{**}(z)$, $\omega^{**}(z)$ they may be represented by Laurent series of the form $\sum_{n=-\infty}^{\infty} \alpha_n z^n$ for $z \in E$ (see Section 1.3). The Laurent series are uniformly convergent in every finite region in E.†

If we now stipulate that the stresses exist and are bounded at infinity, from (2.44) it will be seen that $\Omega'(z)$, $\omega'(z)$ must have $O(1)$ as $|z| \to \infty$. Thus $\Omega^{**}(z)$, $\omega^{**}(z)$ have Laurent expansions of the form

$$\begin{aligned} \Omega^{**}(z) &= (A + \mathrm{i}B)z + \sum_{n=0}^{\infty} \alpha_n z^{-n} \\ \omega^{**}(z) &= (C + \mathrm{i}D)z + \sum_{n=0}^{\infty} \beta_n z^{-n} \end{aligned} \tag{2.60}$$

† There is the possibility that the Laurent series are not convergent for all values of $\arg(z)$ as $|z| \to \infty$.

and from (2.58) and (2.60) the stress field at infinity is given by

$$\tau_{xx} + \tau_{yy} = 4A$$
$$\tau_{xx} - \tau_{yy} + 2\mathrm{i}\tau_{xy} = -2(C - \mathrm{i}D).$$

Hence the stress distribution at infinity must be *uniform*.

If we suppose the principal stresses at infinity are N_1, N_2 where the angle between N_1 and the x-axis is α then

$$\begin{aligned} 4A &= N_1 + N_2 \\ 2(C - \mathrm{i}D) &= -(N_1 - N_2)\mathrm{e}^{2\mathrm{i}\alpha}. \end{aligned} \tag{2.61}$$

These formulae do not involve B which is related to the rotation ω_∞ at infinity

$$\omega_\infty = \mathrm{Im}\left(\frac{\partial D}{\partial z}\right) = (1 + \kappa)B/2\mu. \tag{2.62}$$

The corresponding displacement at infinity is

$$\begin{aligned} 2\mu D = -\frac{\kappa(X + \mathrm{i}Y)}{2\pi(1 + \kappa)}\log(z\bar{z}) + \{(\kappa - 1)A + \mathrm{i}(\kappa + 1)B\}z \\ - (C - iD)\bar{z} + O(1) \end{aligned} \tag{2.63}$$

and is unbounded. We note that even in the case of zero stress and rotation at infinity when $A = B = C = D = 0$ the resultant force $X + \mathrm{i}Y$ over the internal boundaries gives rise to a displacement field of the form

$$2\mu D = -\frac{\kappa(X + \mathrm{i}Y)}{\pi(1 + \kappa)}\log(|z|) \tag{2.64}$$

as $|z| \to \infty$.

Semi-infinite regions

If we suppose the region S^+ is a semi-infinite region bounded externally by the contour C_0, which passes through the finite part of the plane, and internally by the set of holes with bounding contours $C_1, C_2, \ldots, C_n$, then the foregoing discussion breaks down as Laurent's theorem applies to annular regions only and consequently the series representations of (2.60) are not sufficiently general. To make the problem more precise let us assume C_0 divides the plane into two semi-infinite regions and has a positive direction so that S^+ lies to the left of C_0. Further we suppose C_0 tends asymptotically to the lines $\theta = \alpha$, $\theta = \pi - \alpha$ in the positive and negative directions respectively. This situation is illustrated in Figure 2.5. Again it is clear that a radius R_0 can be defined so that $|z| < R_0$ covers the contours $C_1, C_2, \ldots, C_n$, and the discussion leading to equations (2.58)

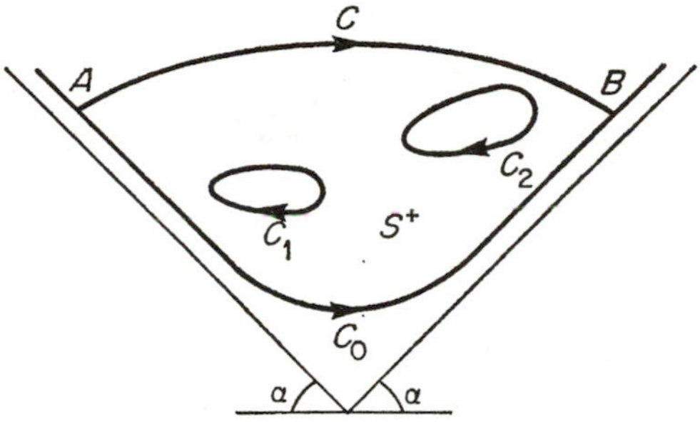

Figure 2.5

and (2.59) still applies. However, in this case the functions $\Omega^{**}(z)$, $\omega^{**}(z)$ are holomorphic and single valued in the region S^+ for $|z| \geqslant R_0$. As this region is approximately the sector $|z| \geqslant R_0$ $(\alpha \leqslant \arg(z) \leqslant \pi - \alpha)$, Laurent's theorem may not be applied.

If we consider the case where the stresses and rotation tend to zero as $|z|$ tends to infinity it may be shown that the potentials have the form

$$\Omega^{**}(z) = \beta \log z + \sum_{n=0}^{\infty} \alpha_n z^{-n}$$

$$\omega^{**}(z) = \beta' \log z + \sum_{n=0}^{\infty} \beta_n z^{-n}$$

where such log terms were excluded from the Laurent expansion. Consequently for large $|z|$ the potentials $\Omega(z)$, $\omega(z)$ have the form

$$\Omega(z) = \gamma_1 \log z + O(1), \qquad \omega(z) = \gamma_2 \log z + O(1). \tag{2.65}$$

Let us suppose the resultant force applied to the contour C_0 is finite, and that the stresses on C_0 have $O(|z|^{-\mu})$, $\mu > 1$ for large $|z|$. Then the resultant force applied to the body S^+ over the contours $C_0, C_1, \ldots, C_n$ is taken to be finite and we shall denote it by P. Let us consider a contour ACB in S^+ joining the points A, B of C_0, which lies sufficiently far from the origin so that the closed contour formed from ACB and the part of C_0 joining B and A contains within it the contours $C_1, C_2, \ldots, C_n$. This is shown in Figure 2.5. Then the resultant force on the material to the left of ACB tends to P as the contour ACB tends to infinity. This resultant force may be simply calculated.

Because of the assumed asymptotic behaviour of C_0, the points A and B which tend to infinity along the negative and positive directions of C_0

may be approximated by putting

$$A = R_1 e^{i(\pi - \alpha)}, \qquad B = R_2 e^{i\alpha}$$

where R_1, R_2 tend to infinity independently. From (2.47) the resultant force over the arc ACB is

$$-i[\Omega(z) + z\,\overline{\Omega'(z)} + \overline{\omega(z)}]_A^B$$
$$= -i\{(\gamma_1 - \bar{\gamma}_2)i(\pi - 2\alpha) + (\gamma_1 + \bar{\gamma}_2)\log(R_2/R_1) - 2i\bar{\gamma}_1 \sin 2\alpha\}.$$

Since this tends to the resultant applied force P as A and B independently tend to infinity, we must conclude

$$\gamma_1 + \bar{\gamma}_2 = 0$$
$$P = [(\gamma_1 - \bar{\gamma}_2)(2\alpha - \pi) + 2\bar{\gamma}_1 \sin 2\alpha]$$

and hence

$$\gamma_1 = -\bar{\gamma}_2 = -\frac{P(\pi - 2\alpha) + \bar{P}\sin 2\alpha}{2[(\pi - 2\alpha)^2 - \sin^2 2\alpha]}.$$

Thus the behaviour of the complex potentials at infinity in this case is quite different from that in the earlier parts of this section. These constants take a particularly simple form if we consider the case $\alpha = 0$ when the region S^+ is essentially a half-plane. In this case

$$\Omega(z) = -\frac{P}{2\pi}\log z + O(1)$$
$$\omega(z) = \frac{\bar{P}}{2\pi}\log z + O(1) \qquad (2.67)$$

for large $|z|$. These expressions should be compared with equations (2.58) and (2.60).

2.11 Boundary conditions for the complex potentials

The results of Sections 2.5, 2.7 enable us to formulate the fundamental boundary value problems of Section 2.3 in terms of the complex potentials.

The boundary conditions of the stress boundary value problem (2.13) take the form

$$\tau_{xx}n_1 + \tau_{xy}n_2 = T_1$$
$$\tau_{xy}n_1 + \tau_{yy}n_2 = T_2$$

where the stress vector (T_1, T_2) is specified on the boundary C of S^+. C is in general the union of the contours $C_0, C_1, \ldots, C_n$. Boundary conditions of this nature were considered at the beginning of Section 2.7 and from (2.47) we find

$$-\mathrm{i}\frac{\mathrm{d}}{\mathrm{d}s}[\Omega(z) + z\,\overline{\Omega'(z)} + \overline{\omega(z)}] = T_1 + \mathrm{i}T_2$$

on C, where s is the arc length measured around the contours of C. We note that an internal contour of S^+ is described in a clockwise direction and the external contour anticlockwise.

On integrating along the contours C we find

$$\Omega(z) + z\,\overline{\Omega'(z)} + \overline{\omega(z)} = f_1 + \mathrm{i}f_2 + \text{constant} \quad (z \in C) \qquad (2.68)$$

where

$$f_1 + \mathrm{i}f_2 = \mathrm{i}\int^z (T_1 + \mathrm{i}T_2)\,\mathrm{d}s$$

this last integral being taken from some fixed point on each contour to an arbitrary point z on the contour keeping the material to the left. Clearly the complex constant in (2.68) may have different values on each contour C_k. Further the term $\Omega(z) + z\,\overline{\Omega'(z)} + \overline{\omega(z)}$ in (2.68) denotes the limit of this expression as z tends to a point of C from the inside of body S^+.

Thus the stress boundary value problem reduces to the determination of the complex potentials $\Omega(z)$, $\omega(z)$ subject to the condition (2.68) on the boundary C of S^+. Muskhelishvili,[43] Section 40 has shown that the stress boundary value problem uniquely determines the stresses in the whole body but that the corresponding displacements are only defined to within a rigid body displacement. Hence, as in Section 2.8, $\Omega(z)$ and $\omega(z)$ are not completely defined and we may replace $\Omega(z)$ by $\Omega(z) + \mathrm{i}Az + \alpha$ and $\omega(z)$ by $\omega(z) + \beta$ without changing the stress distribution. Substituting these expressions in (2.68) changes the left-hand side by a constant $\alpha + \bar{\beta}$. Hence by a suitable choice of $\alpha + \bar{\beta}$ one of the constants on the right-hand side of (2.68) may be assigned arbitrarily, but the remaining constants are not at our disposal and are determined from the conditions of continuity and single-valuedness of the displacements and stresses. This is discussed later in Section 3.10. We note in passing that as the equations of equilibrium were taken as part of our basic equations, solutions to (2.68) for a bounded region can only be obtained provided the resultant force and moment of the applied forces are zero. A second formulation of this problem in which the normal and tangential components of the stress vector are prescribed on the boundary is given in the next section.

If we now examine the displacement boundary value problem of Section 2.3 we find the complex potentials satisfy a similar boundary condition to (2.68). From (2.14)

$$D = u + \mathrm{i}v = g_1 + \mathrm{i}g_2, \quad \text{on } C$$

where $g_1(x, y)$, $g_2(x, y)$ are the given boundary displacements. Using (2.44), this clearly results in the equation

$$\kappa\,\Omega(z) - z\,\overline{\Omega'(z)} - \overline{\omega(z)} = 2\mu(g_1 + ig_2) \tag{2.69}$$

where as in (2.68) the left side must be regarded as the limiting value from the inside of S^+.

Similarly if we formulate the fundamental mixed boundary value problem in terms of the complex potentials it will be seen that the condition (2.68) must hold over part of C while (2.69) holds over the remainder of C. Muskhelishvili,[43] Section 40 has shown that both the displacement and the mixed boundary value problems uniquely determine the displacements (and hence also the stresses) in S^+. A more detailed discussion of these boundary conditions has been given by Muskhelishvili,[43] Section 41.

Although the problems of proving existence and uniqueness of the complex potentials satisfying equations (2.68), or (2.69), or for the mixed problem for both simply connected or multiply connected regions S^+ are of fundamental importance, it is sufficient for the purpose of this monograph to indicate that under certain assumptions solutions to the above problems may be shown to exist and be unique. Papers dealing with these problems have been published by several authors including Muskhelishvili, Mikhlin and Sherman and references have been given by Sokolnikoff,[59] Section 74 and Muskhelishvili,[43] Sections 40, 42, see also the review by Teodorescu.[63] In particular the condition of regularity assumed by Muskhelishvili,[43] Section 42 to prove uniqueness is that the expression $\Omega(z) + z\,\overline{\Omega'(z)} + \overline{\omega(z)}$ is continuous up to the boundary C of the region S^+. The continuity of this expression is equivalent to the statement that the resultant force over an arbitrarily placed arc in the body tends to zero as the length of the arc tends to zero. Hence in the case of a point force acting on the boundary C the regularity condition is violated. However Tiffen[64] has shown that the solutions obtained are unique even in the case of isolated point forces both in the body S^+ and on the boundary C.

2.12 Stresses in curvilinear coordinates

Suppose we consider a system of orthogonal curvilinear coordinates (ξ, η) in the z-plane. An element of arc of the curve $\xi = \xi_0$ is perpendicular to the direction in which ξ increases. Let us denote the normal stress across this element by $\tau_{\xi\xi}$ and the shear stress by $\tau_{\xi\eta}$, see Figure 2.6.

Similarly across an element of arc of the curve $\eta = \eta_0$ the normal stress is $\tau_{\eta\eta}$ and the shear stress $\tau_{\eta\xi}$. As there are no couple stresses $\tau_{\xi\eta} = \tau_{\eta\xi}$. We denote the displacements along the ξ direction and the η direction by

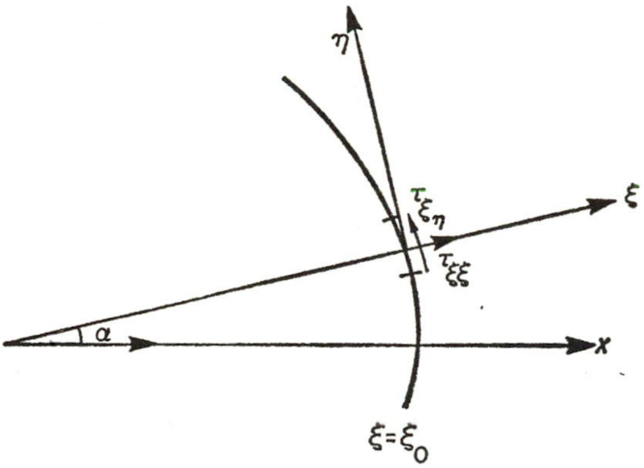

Figure 2.6

u_ξ and u_η respectively and suppose the tangent to the ξ curve makes an angle α with the x-axis.

Taking components of the displacement vector along the ξ and η directions we find

$$u_\xi = u\cos\alpha + v\sin\alpha, \qquad u_\eta = -u\sin\alpha + v\cos\alpha$$

so that

$$u_\xi + \mathrm{i}u_\eta = (u + \mathrm{i}v)\mathrm{e}^{-\mathrm{i}\alpha} \tag{2.70}$$

Similarly on examining the forces across elements of arc perpendicular to the ξ and η directions the normal and shear stresses may be found. In terms of the stress combinations the transformation laws are

$$\begin{aligned} \tau_{\xi\xi} + \tau_{\eta\eta} &= \tau_{xx} + \tau_{yy} = \Theta \\ \tau_{\xi\xi} - \tau_{\eta\eta} + 2\mathrm{i}\tau_{\xi\eta} &= (\tau_{xx} - \tau_{yy} + 2\mathrm{i}\tau_{xy})\mathrm{e}^{-2\mathrm{i}\alpha} = \Phi\mathrm{e}^{-2\mathrm{i}\alpha} \end{aligned} \tag{2.71}$$

Thus Θ, the sum of the normal stress across two perpendicular directions, is an invariant under the rotation of the axes. If we now substitute for the stresses and displacements in terms of the complex potentials of (2.44) we find

$$\begin{aligned} \tau_{\xi\xi} + \tau_{\eta\eta} &= 2\{\Omega'(z) + \overline{\Omega'(z)}\} \\ \tau_{\xi\xi} - \tau_{\eta\eta} + 2\mathrm{i}\tau_{\xi\eta} &= -2\{z\,\overline{\Omega''(z)} + \overline{\omega'(z)}\}\mathrm{e}^{-2\mathrm{i}\alpha} \\ 2\mu(u_\xi + \mathrm{i}u_\eta) &= \{\kappa\,\Omega(z) - z\,\overline{\Omega'(z)} - \overline{\omega(z)}\}\mathrm{e}^{-\mathrm{i}\alpha} \end{aligned} \tag{2.72}$$

and hence

$$\tau_{\xi\xi} + \mathrm{i}\tau_{\xi\eta} = \Omega'(z) + \overline{\Omega'(z)} - \{z\,\overline{\Omega''(z)} + \overline{\omega'(z)}\}\mathrm{e}^{-2\mathrm{i}\alpha} \tag{2.73}$$

In particular if ξ, η are polar coordinates r, θ, then $\alpha = \theta$ and (2.72), (2.73) yield the useful formulae

$$\begin{aligned} \tau_{rr} + \tau_{\theta\theta} &= 2\{\Omega'(z) + \overline{\Omega'(z)}\} \\ \tau_{rr} - \tau_{\theta\theta} + 2\mathrm{i}\tau_{r\theta} &= -2\left\{\bar{z}\,\overline{\Omega''(z)} + \frac{\bar{z}}{z}\,\overline{\omega'(z)}\right\} \end{aligned} \tag{2.74}$$

on replacing $\mathrm{e}^{-2\mathrm{i}\theta}$ by $\bar{z}/z$. Also

$$2\mu(u_r + \mathrm{i}u_\theta) = \mathrm{e}^{-\mathrm{i}\theta}\{\kappa\,\Omega(z) - z\,\overline{\Omega'(z)} - \overline{\omega(z)}\}. \tag{2.75}$$

In the previous section when considering the stress boundary value problem we prescribed the applied force over the boundary C in terms of its cartesian components T_1, T_2. It is equally possible however to specify the normal N and shear stress T over the boundary C. If the curvilinear coordinate system ξ, η is chosen so that the part of the boundary C over which the stress is specified coincides with $\xi = \xi_0$ then the boundary condition becomes

$$\tau_{\xi\xi} + \mathrm{i}\tau_{\xi\eta} = N + \mathrm{i}T$$

on part of $\xi = \xi_0$.

Hence from (2.73) the complex potentials satisfy a boundary condition of the form

$$\Omega'(z) + \overline{\Omega'(z)} - \mathrm{e}^{-2\mathrm{i}\alpha}\{z\,\overline{\Omega''(z)} + \overline{\omega'(z)}\} = N + \mathrm{i}T \quad (z \in C)$$

where α is the angle of inclination of the external normal to C. This form for the boundary conditions is often convenient since even in multiply connected regions $\Omega'(z)$ and $\omega'(z)$ are single-valued functions (see Section 2.9).

Alternatively, from (2.47), the stress boundary value problem may be expressed in terms of the resultant force $R(t)$ over an arbitrary arc of the Boundary C from $z = t_0$ to $z = t$ as

$$\begin{aligned} [\Omega(z) + z\,\overline{\Omega'(z)} + \overline{\omega(z)}]_{t_0}^{t} &= \mathrm{i}\,R(t) \\ &= \mathrm{i}\int_{z=t_0}^{z=t} (\tau_{\xi\xi} + \mathrm{i}\tau_{\xi\eta})\mathrm{e}^{\mathrm{i}\alpha}\,\mathrm{d}s \\ &= \int_{t_0}^{t} (\tau_{\xi\xi} + \mathrm{i}\tau_{\xi\eta})\mathrm{d}t \quad (t \in C) \end{aligned} \tag{2.76}$$

where the direction of integration keeps S^+ to the left.

For circular regions the above formulae allow many simple results to be obtained. If the region S^+ has a more complicated shape the relationship between α and the coordinates z, $\bar{z}$ is less simple. However a powerful method of dealing with a wide class of such problems exists. If it is possible to find a *conformal transformation* which will transform the region in question into one of more simple shape, for example a circular region or a half-plane, then it might be possible to solve the transformed boundary problem in this 'image' region. This method is investigated in Chapter 5.

Appendix

This appendix and the first three examples give two further methods of derivation of the general solution of the equations of two-dimensional elasticity. Both methods are based on the Helmholtz representation† of a two-dimensional vector **u** in the form

$$u_\alpha = \phi_{,\alpha} + e_{\alpha\beta}\psi_{,\beta} \tag{2.77}$$

where $e_{\alpha\beta}$ is the two-dimensional alternating tensor ($e_{11} = e_{22} = 0, e_{12} = 1, e_{21} = -1$) and ϕ and ψ are single-valued functions of position. We note $u_1 = \phi_{,1} + \psi_{,2}$, $u_2 = \phi_{,2} - \psi_{,1}$ so that

$$\begin{aligned} \nabla_1^2\phi &= u_{\alpha,\alpha} = \nabla_1 \cdot \mathbf{u} \\ \nabla_1^2\psi &= e_{\alpha\beta}u_{\alpha,\beta}. \end{aligned} \tag{2.78}$$

We now use the representation (2.77) to derive a general solution of Navier's equations (2.28) namely

$$(\lambda + \mu)\nabla_1(\nabla_1 \cdot \mathbf{u}) + \mu\nabla_1^2\mathbf{u} + \rho\mathbf{F} = 0.$$

Substituting from (2.78) we find

$$\nabla_1^2\{(\lambda + \mu)\nabla_1\phi + \mu\mathbf{u}\} + \rho\mathbf{F} = 0$$

so that on putting

$$(\lambda + \mu)\nabla_1\phi + \mu\mathbf{u} = \frac{(\lambda + 2\mu)}{\lambda + \mu}\boldsymbol{\theta} \tag{2.79}$$

the vector function $\boldsymbol{\theta}$ satisfies the Poisson equation

$$\nabla_1^2\boldsymbol{\theta} = -\frac{\rho(\lambda + \mu)}{(\lambda + 2\mu)}\mathbf{F}. \tag{2.80}$$

Assuming $\boldsymbol{\theta}$ can be found from this equation, **u** may be determined from (2.79) once ϕ is known. On taking the divergence of (2.79)

$$(\lambda + \mu)\nabla_1^2\phi + \mu\nabla_1 \cdot \mathbf{u} = \frac{(\lambda + 2\mu)}{(\lambda + \mu)}\nabla_1 \cdot \boldsymbol{\theta}$$

and using (2.78), ϕ satisfies the relation

$$\nabla_1^2\phi = \frac{1}{(\lambda + \mu)}\nabla_1 \cdot \boldsymbol{\theta}. \tag{2.81}$$

† A proof of this result may be found in Epstein.[13]

This equation has a general integral of the form

$$\phi = \frac{1}{2(\lambda + \mu)}(x_\alpha \theta_\alpha + \phi_0) \tag{2.82}$$

where

$$\begin{aligned} \nabla_1^2 \phi_0 &= -x_\alpha \nabla_1^2 \theta_\alpha \\ &= \frac{\rho(\lambda + \mu)}{(\lambda + 2\mu)} x_\alpha F_\alpha \end{aligned} \tag{2.83}$$

Hence from (2.79) and (2.82)

$$\begin{aligned} 2\mu u_\alpha &= \left\{\frac{2(\lambda + 2\mu)}{\lambda + \mu} - 1\right\}\theta_\alpha - x_\beta \theta_{\beta,\alpha} - \phi_{0,\alpha} \\ &= \kappa \theta_\alpha - x_\beta \theta_{\beta,\alpha} - \phi_{0,\alpha} \end{aligned} \tag{2.84}$$

on using the notation of (2.39). This solution of Navier's equations in terms of three potential functions θ_1, θ_2 and ϕ_0 which satisfy the Poisson equations (2.80) and (2.83) is the analogue in two-dimensional elasticity of the solution of the corresponding Navier's equations of three-dimensional linear elasticity derived by Papkovich and Neuber†. The relation between this solution and the complex variable representation derived in Section 2.5 is examined in the first example.

Examples (2) and (3) indicate an alternative method of derivation of the general solution which follows from an examination of the system of equations satisfied by the stresses. In this case a general representation is found for the stresses in the body in terms of a multiple-valued potential function (the Airy stress function) which may then be integrated to determine the corresponding displacement field. This approach has been used by Muskhelishvili,[43] Chapter V, in deriving the complex variable formulation of the general solution. The corresponding stress function representations in three-dimensional linear elasticity have been examined by Rieder,[49] Gurtin,[23,24] and Mindlin.[42]

Examples on Chapter 2

1. In the case when the body force $\mathbf{F} = 0$ show that the complex displacement $D = u_1 + iu_2$ may be expressed in terms of two arbitrary functions of a complex variable by representing the harmonic function θ_1 in (2.84) as the real part of a holomorphic function and θ_2, ϕ_0 as imaginary parts of holomorphic functions. Hence derive formulae (2.44).

† A derivation of the Papkovich–Neuber representation and a discussion of this and other representations is given in Sokolnikoff[59] Chapter 6. See also Mindlin[42] and Love[38] Chapter IX.

2. The basic equations of two-dimensional elasticity when expressed in terms of the stresses are the equations of equilibrium and the equation of compatibility namely

$$\tau_{\alpha\beta,\beta} + \rho F_\alpha = 0$$

$$(\kappa + 1)\nabla_1^2 \tau_{\alpha\alpha} + 4\rho F_{\alpha,\alpha} = 0.$$

When $\mathbf{F} = 0$ show that the general solution of the equilibrium equations can be expressed in the form

$$\begin{aligned} \tau_{11} &= A_{,22} + \psi_{,12} \\ \tau_{22} &= A_{,11} - \psi_{,12} \\ \tau_{12} &= -A_{,12} - \psi_{,11} \end{aligned} \tag{2.85}$$

where $\nabla_1^2\psi = 0$, by applying the Helmholtz decomposition (2.77) to each of the vectors (τ_{11}, τ_{12}) and (τ_{21}, τ_{22}) in the equilibrium equations.

On substituting in the compatibility equations show that $\nabla_1^4 A = 0$ for $\mathbf{F} = 0$. The single-valued function A is the Airy stress function.†

3. On observing that a harmonic function ϕ may be expressed as the real or imaginary part of a complex function holomorphic in S^+, show that the stress field in (2.85) must have the form

$$\tau_{11} + \tau_{22} = 2\{\Omega'(z) + \overline{\Omega'(z)}\}$$

$$\tau_{11} - \tau_{22} + 2i\tau_{12} = -2\{z\,\overline{\Omega''(z)} + \overline{\omega'(z)}\}$$

where $\Omega'(z)$, $\omega'(z)$ are holomorphic in S^+.

Hence use the stress displacement relations (2.43) to determine the form of the complex displacement D in terms of $\Omega(z)$, $\omega(z)$.

4. Prove that $\Omega(z)$, $\omega(z)$, and hence the stress field, are independent of the elastic constants in the stress boundary value problem for a simply connected region.

5. Find the stress and displacement fields corresponding to the complex potentials

$$\Omega(z) = \alpha + iAz, \qquad \omega(z) = \beta$$

where α and β are complex constants and A is real. Interpret this displacement field.

† Note that for a simply connected region S^+ the effect of the harmonic function ψ in (2.85) may be included in the biharmonic function A. This is not the case for a multiply connected region S^+ where a single-valued Airy stress function can only correspond to a stress field in which the resultant force over each internal boundary of S^+ is zero, ψ however can be dispensed with if A is allowed to be multiple-valued. For further information see Rieder[49] and Gurtin.[24]

6. Show that the complex potentials

$$\Omega(z) = \alpha z, \qquad \omega(z) = \beta z$$

correspond to a uniform stress field. Choose α and β so that

$$\tau_{xx} = X, \qquad \tau_{xy} = S, \qquad \tau_{yy} = Y.$$

Why is $\mathrm{Im}(\alpha)$ not specified? For this uniform stress field use the relation (2.73) to find the stresses across the lines x = constant, y = constant and $x + y$ = constant.

7. Use (2.74) to find the stress distribution on the circle $r = a$ corresponding to the complex potentials

$$\Omega(z) = \alpha z^n, \qquad \omega(z) = \beta z^n.$$

Hence determine the complex potentials for the following problems

(*a*) A circular disc $|z| < b$ under the stress distribution $\tau_{rr} + \mathrm{i}\tau_{r\theta} = -p_2 + \mathrm{i}s_2$ on $|z| = b$. Why must $s_2 = 0$?

(*b*) An infinite medium $|z| > a$ with zero stress and rotation at infinity subject to the stress boundary condition

$$\tau_{rr} + \mathrm{i}\tau_{r\theta} = -p_1 + \mathrm{i}s_1$$

on the hole $|z| = a$.

(*c*) A circular annulus $a < |z| < b$ with the stress boundary conditions

$$\begin{aligned} \tau_{rr} + \mathrm{i}\tau_{r\theta} &= -p_1 + \mathrm{i}s_1 \qquad \text{on } |z| = a \\ &= -p_2 + \mathrm{i}s_2 \qquad \text{on } |z| = b. \end{aligned}$$

What is the physical interpretation of the relation between s_1 and s_2?

8. Complex potentials of the form

$$\Omega(z) = -\frac{X + \mathrm{i}Y}{2\pi(1 + \kappa)} \log z, \qquad \omega(z) = \frac{\kappa(X - \mathrm{i}Y)}{2\pi(1 + \kappa)} \log z$$

have been found to arise in problems where the body contains a hole surrounding the origin (see Section 2.9). Show that the corresponding complex displacement D is unbounded as $|z| \to \infty$ and as $|z| \to 0$. Also by use of (2.47) show that the resultant force across any contour surrounding the origin is $X + \mathrm{i}Y$. Find the stress distribution on the circle $r = a$.

9. Show that the complex potentials

$$\Omega(z) = -\frac{X + \mathrm{i}Y}{2\pi(1 + \kappa)} \log z, \qquad \omega(z) = \frac{\kappa(X - \mathrm{i}Y)}{2\pi(1 + \kappa)} \log z + \frac{(X + \mathrm{i}Y)a^2}{2\pi(1 + \kappa)z^2}$$

yield the solution for the infinite medium $|z| > a$ with zero stress and rotation at infinity under the action of the stress distribution

$$\tau_{rr} + \mathrm{i}\tau_{r\theta} = -\frac{1}{2\pi a}(X + \mathrm{i}Y)\mathrm{e}^{-\mathrm{i}\theta}$$

on $|z| = a$. Note that the resultant force over the hole is $X + \mathrm{i}Y$ and that as $a \to 0$ the potentials given above tend to those of example 8. The limit as $a \to 0$ of this particular physical problem is used to define a point force $X + \mathrm{i}Y$ acting at the origin (see Sections 3.2, 4.4).

10. Determine the stress field in a uniform disc $|z| < a$ rotating with uniform angular velocity about its centre, ensuring that the edge $|z| = a$ is unstressed.

3

PLANE AND HALF-PLANE PROBLEMS

3.1 Infinite plane

We begin this chapter by examining the stress distribution that can exist in an infinite plane. It is clear from the solutions (2.44) to the basic equations of elasticity that any choice of the complex potentials $\Omega(z)$, $\omega(z)$ will yield a stress field and the corresponding displacement field. In general the problem is to determine $\Omega(z)$, $\omega(z)$ subject to certain conditions of the type formulated in Section 2.12 on the boundary C of the region S^+ occupied by the body. Clearly if the region S^+ is the whole plane, that is the body does not contain any holes, then by the results of Section 2.9 $\Omega(z)$, $\omega(z)$ must be single-valued holomorphic functions in the whole plane except possibly at infinity, and hence have Laurent expansions of the form

$$\Omega(z) = \sum_{n=0}^{\infty} \alpha_n z^n, \qquad \omega(z) = \sum_{n=0}^{\infty} \beta_n z^n.$$

Thus any deformation of the plane is accompanied by infinite displacements at infinity. If, as in Section 2.10, we assume the stresses exist and are bounded at infinity then $\Omega(z)$, $\omega(z)$ are linear functions of z and consequently the stresses are uniform at infinity. If we suppose

$$\tau_{xx} = T_1, \qquad \tau_{xy} = S, \qquad \tau_{yy} = T_2 \tag{3.1}$$

at infinity then

$$\Omega(z) = \tfrac{1}{4}(T_1 + T_2)z, \qquad \omega(z) = -\tfrac{1}{2}(T_1 - T_2 - 2iS)z \tag{3.2}$$

and hence the stresses are uniform throughout the whole plane and are given by (3.1). Thus the most general deformation the infinite plane can undergo with bounded stresses at infinity is a combination of uniform extensions due to the stresses T_1 and T_2 and a uniform shear. This situation is radically altered if point forces can occur in the interior of S^+, or if S^+ has a boundary other than the point at infinity.

3.2 Point forces and moments

Suppose we consider an infinite plate containing a single hole bounded by the contour C_1. Then if the origin lies in the interior of the hole and a resultant force $X + \mathrm{i}Y$ acts over C_1 we have shown (2.57) that the complex potentials have the form

$$\Omega(z) = -\frac{X + \mathrm{i}Y}{2\pi(1 + \kappa)}\log z + \Omega^*(z)$$
$$\omega(z) = \frac{\kappa(X - \mathrm{i}Y)}{2\pi(1 + \kappa)}\log z + \omega^*(z)$$

where $\Omega^*(z)$, $\omega^*(z)$ are holomorphic and single valued in the body. If the stress and rotation are zero at infinity then $\Omega^*(z)$, $\omega^*(z)$ have expansions of the form $\sum_{n=1}^{\infty} \alpha_n z^{-n}$ for large $|z|$. Hence the terms which dominate the complex potentials at a large distance from the hole are the terms in $\log z$. This suggests that a point force $X + \mathrm{i}Y$ acting at the origin may be represented by the potentials

$$\Omega(z) = -\frac{X + \mathrm{i}Y}{2\pi(1 + \kappa)}\log z, \qquad \omega(z) = \frac{\kappa(X - \mathrm{i}Y)}{2\pi(1 + \kappa)}\log z. \tag{3.3}$$

It is simple to confirm that for any contour C surrounding $z = 0$ the resultant force over the contour is $X + \mathrm{i}Y$ and the resultant moment is zero using the results of Section 2.7. However it is apparent that additional potentials may be added to those of (3.3) and, provided the additional potentials have no singularities for $z \neq 0$ and leave the resultant force and moment at the origin unchanged, an alternative type of point force will result. For example, possible additional potentials are $\Omega(z) = Az^{-2}$, $\omega(z) = Bz^{-2}$. This apparent arbitrariness of a point force can only be resolved when it is defined as the limit of a given stress distribution over a given hole. For the time being we shall define a point force by equations (3.3). A particular example of such a limit has been given in Chapter 2, example 9.

In a similar manner we can define a point moment M at the origin by the potentials

$$\Omega(z) = 0, \qquad \omega(z) = \mathrm{i}M/2\pi z \tag{3.4}$$

and it will be seen that these potentials do not induce a resultant point force at the origin. Corresponding to both sets of potentials (3.3) and (3.4) the displacement field due to a point force $X + \mathrm{i}Y$ and moment M at the origin is

$$4\pi\mu(1 + \kappa)D = -\kappa(X + \mathrm{i}Y)\log(z\bar{z}) + (X - \mathrm{i}Y)(z/\bar{z}) + \mathrm{i}(1 + \kappa)(M/\bar{z}) \tag{3.5}$$

and is unbounded at the origin and infinity. It is clear that a singularity at the origin of this nature requires some explanation. The potentials (3.3) and (3.4) correspond to particular stress distributions over a small hole C_1 about the origin and have been constructed under the assumption that the point z is sufficiently far from the origin that the fine structure of the stress distribution on C_1 does not affect the stress or displacement fields at the point z. Consequently potentials (3.3) and (3.4) yield a meaningful physical picture only at points which are 'distant' from the origin. A careful discussion of these properties involves consideration of Saint Venant's Principle and the reader is referred to the recent work of Sternberg,[60] Toupin[68] and Knowles,[29] for further details. The presence of these singularities is somewhat analogous to the situation in hydrodynamics where infinite velocities and pressures or tensions occur at sources, sinks or vortices.

If the point force $X + \mathrm{i}Y$ and moment M act at a point z_0 the corresponding potentials are

$$\begin{aligned} \Omega(z) &= -\frac{X+\mathrm{i}Y}{2\pi(1+\kappa)}\log(z-z_0) \\ \omega(z) &= \frac{\kappa(X-\mathrm{i}Y)}{2\pi(1+\kappa)}\log(z-z_0) + \frac{X+\mathrm{i}Y}{2\pi(1+\kappa)}\cdot\frac{\bar{z}_0}{(z-z_0)} + \frac{\mathrm{i}M}{2\pi(z-z_0)}. \end{aligned} \tag{3.6}$$

3.3 Continuation of the complex potentials

The process of analytic continuation, combined with the Cauchy integral representations given in Chapter 1 for sectionally holomorphic functions with given discontinuities on a set of arcs or contours, forms one of the most powerful techniques for the solution of boundary value problems in two-dimensional elasticity. This method owes its foundation to Muskhelishvili and is used extensively in his two books.[43, 44] In the remainder of this chapter we shall illustrate the use of this method in the solution of some typical plane or half-plane problems.†

We consider first a material occupying the upper half-plane $y > 0$ and denote this region by S^+, using the notation S^- for the half-plane $y < 0$. Now the stress and displacement fields in the body are determined entirely by the values of the complex potentials in S^+ and their boundary values on $y = 0+$ from equations (2.44). Consequently the complex potentials $\Omega(z)$, $\omega(z)$ may be specified arbitrarily in S^- without affecting the stress and displacement fields. This fact enables us to simplify the solution of

† For further references see Muskhelishvili[43] Chapters 16, 19, 20, Green and Zerna[21] Chapter 8, Milne-Thomson.[41]

certain problems by replacing the definition of one of the potentials—say $\omega(z)$—in S^+ by the definition of the other potential $\Omega(z)$ in S^-. This being done the elastic state of the body may then be expressed in terms of a single potential defined over the whole plane. The method of continuation of $\Omega(z)$ from S^+ into S^- is normally chosen to make the form of the resulting boundary conditions as simple as possible. We now consider the most common continuations.

From (2.44) the normal stress has the form

$$\tau_{yy} - i\tau_{xy} = \Omega'(z) + \overline{\Omega'(z)} + z\overline{\Omega''(z)} + \overline{\omega'(z)}.$$

Let us suppose that a region L of the surface $y = 0$ of the half-plane S^+ is unstressed so that

$$\lim_{y\to 0+} \{\Omega'(z) + \overline{\Omega'(z)} + z\,\overline{\Omega''(z)} + \overline{\omega'(z)}\} = 0$$

that is

$$\Omega'^+(x) + \overline{\Omega'^+(x)} + x\,\overline{\Omega''^+(x)} + \overline{\omega'^+(x)} = 0 \quad (x \in L) \tag{3.7}$$

using the notation $\lim_{y\to 0+} \Omega(z) = \Omega^+(x)$. This boundary condition may be written in an alternative form by expressing it in terms of the associated functions $\overline{\Omega(\bar{z})}$ and $\overline{\omega(\bar{z})}$ which are holomorphic in S^- (see Section 1.3). Now since

$$\lim_{y\to 0-} \overline{\Omega(\bar{z})} = \lim_{y\to 0+} \overline{\Omega(z)} = \overline{\Omega^+(x)}$$

the boundary condition becomes

$$\Omega'^+(x) = -\lim_{y\to 0-} \{\overline{\Omega'(\bar{z})} + z\,\overline{\Omega''(\bar{z})} + \overline{\omega'(\bar{z})}\} \quad (x \in L).$$

This condition is equivalent to the statement that the holomorphic functions $\Omega'(z)$ for $z \in S^+$ and $-\overline{\Omega'(\bar{z})} - z\,\overline{\Omega''(\bar{z})} - \overline{\omega'(\bar{z})}$ for $z \in S^-$ have equal boundary values on the region L of $y = 0$, and hence, by the continuation theorem (Section 1.3), continue each other analytically across the region L of $y = 0$. In this situation it is natural to *extend* the definition of $\Omega'(z)$ from S^+ into S^- by putting

$$\begin{aligned} \Omega'(z) &= \Omega'(z) \quad (z \in S^+) \\ \Omega'(z) &= -\overline{\Omega'(\bar{z})} - z\,\overline{\Omega''(\bar{z})} - \overline{\omega'(\bar{z})} \quad (z \in S^-). \end{aligned}$$

Hence $\Omega'(z)$ is continued analytically from S^+ into S^- across the part L of $y = 0$ which is unstressed. Clearly these equations may be integrated to give

$$\begin{aligned} \Omega(z) &= \Omega(z) & (z \in S^+) \\ \Omega(z) &= -z\,\overline{\Omega'(\bar{z})} - \overline{\omega(\bar{z})} & (z \in S^-) \end{aligned} \tag{3.8}$$

A result which could be obtained by applying the above arguments to the resultant force (2.47) across a part of L.

These equations may be rearranged to express $\omega(z)$, $z \in S^+$ in terms of $\Omega(z)$ defined in both S^+ and S^- giving

$$\omega(z) = -\overline{\Omega(\bar{z})} - z\,\Omega'(z) \quad (z \in S^+). \tag{3.9}$$

Substituting in the stress and displacement field from (2.44) we find

$$\begin{aligned} 2\mu D &= \kappa\,\Omega(z) + \Omega(\bar{z}) + (\bar{z} - z)\,\overline{\Omega'(z)} \\ \tau_{xx} + \tau_{yy} &= 2\{\Omega'(z) + \overline{\Omega'(z)}\} \\ \tau_{yy} - \mathrm{i}\tau_{xy} &= \Omega'(z) - \Omega'(\bar{z}) + (z - \bar{z})\,\overline{\Omega''(z)} \end{aligned} \tag{3.10}$$

these formulae holding for $z \in S^+$, where $\Omega(z)$ is holomorphic in both S^+ and S^-.

At this stage it is of interest to investigate the conditions satisfied by $\Omega(z)$ as $|z| \to \infty$. From Section 2.10, if the stress and rotation are zero at infinity we have shown

$$\Omega'(z) = -\frac{P}{2\pi z} + O\left(\frac{1}{z^2}\right) \quad (z \in S^+)$$

and

$$\omega'(z) = \frac{\bar{P}}{2\pi z} + O\left(\frac{1}{z^2}\right)$$

hence from (3.8)

$$\Omega'(z) = -\frac{P}{2\pi z} + O\left(\frac{1}{z^2}\right) \tag{3.11}$$

for all z as $|z| \to \infty$ where P is the resultant force applied to the material in S^+.

Assuming $\lim_{y \to 0+} y\,\Omega''(z) = 0$, which can only be confirmed after $\Omega(z)$ is obtained in a particular problem, we see that the displacement and the normal stress on $y = 0+$ are given by

$$\begin{aligned} \tau_{yy} - \mathrm{i}\tau_{xy} &= \Omega'^{+}(x) - \Omega'^{-}(x) \\ 2\mu(u + \mathrm{i}v) &= \kappa\,\Omega^{+}(x) + \Omega^{-}(x) \end{aligned}$$

and (3.12)

$$2\mu\left(\frac{\partial u}{\partial x} + \mathrm{i}\frac{\partial v}{\partial x}\right) = \kappa\,\Omega'^{+}(x) + \Omega'^{-}(x).$$

We note that it is not necessary to define the continuation by means of the homogeneous boundary conditions (3.7) since we can regard the relation (3.9) between $\omega(z)$ and $\Omega(z)$ merely as a convenient representation for $\omega(z)$ which loses no generality and yields the simple boundary values (3.12). We shall refer to the above continuation as the stress continuation.

Clearly alternative methods of extension of the complex potentials could have been adopted. In particular, if we choose to make the extension across portions of the boundary $y = 0$ over which the displacement D is zero, we see that the appropriate continuation of $\Omega(z)$ into the region S^- is

$$\Omega(z) = \frac{1}{\kappa}\{z\,\overline{\Omega'(\bar{z})} + \overline{\omega(\bar{z})}\} \quad (z \in S^-). \tag{3.13}$$

However these two methods of continuation† are really equivalent and we shall in the following restrict our attention to formulae (3.9) and (3.10).

3.4 Stress boundary value problem for a half-plane

The boundary conditions for the stress boundary value problem for a material occupying the half-plane S^+ are, from Section 2.11,

$$\tau_{yy} - \mathrm{i}\tau_{xy} = -\{p(x) + \mathrm{i}s(x)\}, \quad \text{on } y = 0 \tag{3.14}$$

where $p(x)$ is the normal pressure and $s(x)$ is the shear stress applied to the boundary. If we use the stress continuation method outlined above in (3.9) and (3.10) we see that the stress and displacement fields in S^+ may be expressed in terms of a single potential $\Omega(z)$, holomorphic in S^+ and S^- and such that,

$$\Omega'^+(x) - \Omega'^-(x) = -\{p(x) + \mathrm{i}s(x)\} \quad \text{on } y = 0. \tag{3.15}$$

If we suppose further that a finite resultant force P acts on $y = 0$ and the stress and rotation are zero at infinity then from (3.11) $\Omega'(z) = O\left(\frac{1}{z}\right)$ as $|z| \to \infty$. This type of boundary value problem has been considered in Section 1.6 and we see the unique solution is

$$\Omega'(z) = -\frac{1}{2\pi\mathrm{i}}\int_{-\infty}^{\infty} \frac{p(t) + \mathrm{i}s(t)\,\mathrm{d}t}{t - z} \tag{3.16}$$

where it is sufficient‡ to assume $p(t) + \mathrm{i}s(t)$ satisfies the Hölder condition of Section 1.5 for all finite t and the resultant force

$$P = \mathrm{i}\int_{-\infty}^{\infty} p(t) + \mathrm{i}s(t)\,\mathrm{d}t$$

† Another useful continuation is illustrated in Example 7.
‡ Muskhelishvili,[43] p. 394, note 1.

is bounded. In fact these conditions may be relaxed to allow $p(x)$, $s(x)$ finite discontinuities at a finite number of points on $y = 0$.

Several alternative methods of solution of this problem are available and are discussed in Section 3.11.

As a particular example let us consider the effect of a uniform pressure p and shear s over the region $|x| \leqslant a$, the remainder of the boundary $y = 0$ being unstressed. Then

$$\Omega'(z) = -\frac{p + \mathrm{i}s}{2\pi \mathrm{i}}\int_{-a}^{a}\frac{\mathrm{d}t}{t - z}$$

$$= -\frac{p + \mathrm{i}s}{2\pi \mathrm{i}}\log\left(\frac{z - a}{z + a}\right) \tag{3.17}$$

from which the stress and displacement fields may be evaluated. In particular if we put

$$z - a = R_1 \mathrm{e}^{\mathrm{i}\theta_1}, \qquad z + a = R_2 \mathrm{e}^{\mathrm{i}\theta_2}$$

see Figure 3.1, then

$$\tau_{yy} - \mathrm{i}\tau_{xy} = -\frac{(p + \mathrm{i}s)}{\pi}(\theta_1 - \theta_2) + \frac{(p - \mathrm{i}s)}{2\pi \mathrm{i}}[\mathrm{e}^{2\mathrm{i}\theta_1} - \mathrm{e}^{2\mathrm{i}\theta_2}] \tag{3.18}$$

and the corresponding displacement field has an equally simple form.

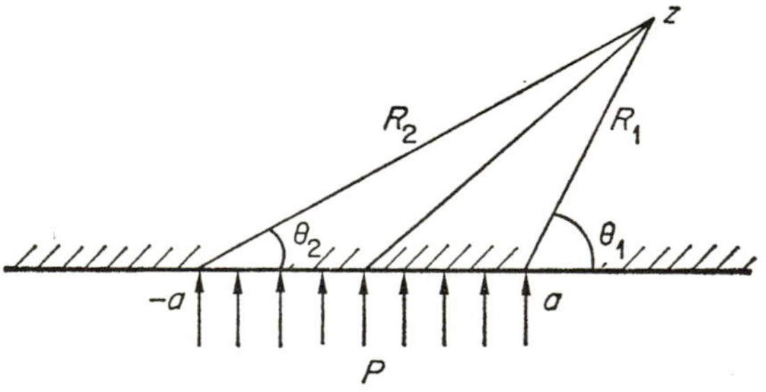

Figure 3.1

These results may also be used to determine the effect of a point force situated on the boundary of the half-plane. If we suppose that as $a \to 0$, p and s tend to infinity in such a way that

$$2ap \to Y, \qquad 2as \to -X$$

where X and Y are constants, then taking the limit as $a \to 0$ in (3.17) we find

$$\Omega'(z) = -\frac{(X + \mathrm{i}Y)}{2\pi z}. \tag{3.19}$$

If we now calculate the stress in the half-plane S^+ it is natural to use polar coordinates and from equations (2.74) we find

$$\tau_{rr} + \tau_{\theta\theta} = \frac{2(X\cos\theta + Y\sin\theta)}{\pi r} = \tau_{rr} - \tau_{\theta\theta} + 2\mathrm{i}\tau_{r\theta}. \quad (3.20)$$

Thus a concentrated force $X + \mathrm{i}Y$ on the boundary of a half-plane produces a purely radial stress

$$\tau_{rr} = -\frac{2(X\cos\theta + Y\sin\theta)}{\pi r}$$

in the half-plane. As might be expected a continuous distribution of point forces on $y = 0$ converts a sum of terms of the form (3.19) into the integral (3.16). (3.19) is the Green's function for the stress boundary value problem for a half-plane.

3.5 Point forces and moments in a half-plane

In some cases it is possible to use the method of continuation to solve a particular problem directly without recourse to the Cauchy integral representations used in the last section. To illustrate this we consider a system of point forces and moments located in S^+ which when acting on a body occupying the whole plane are described by the complex potentials $\Omega_0(z)$, $\omega_0(z)$. It will be seen that $\Omega_0(z)$, $\omega_0(z)$ are sums of terms of the form (3.6), having singularities only in S^+, and are defined for all z. Let us now suppose these point forces and moments act on a body occupying the half-plane S^+, where the surface $y = 0$ is unstressed. It is clear that these forces induce stresses on $y = 0$ of the form

$$\tau_{yy} - \mathrm{i}\tau_{xy} = \Omega_0'(x) + \overline{\Omega_0'(x)} + x\,\overline{\Omega_0''(x)} + \overline{\omega_0'(x)}. \quad (3.21)$$

We must now determine an additional stress field which will remove the stresses (3.21) on $y = 0$ and not give rise to further singularities in S^+. If we denote this additional field by the suffix 1 and represent it by equations (3.10) in terms of a single function $\Omega_1(z)$, holomorphic in S^+ and S^- then $\Omega_1'(z)$ must satisfy the boundary conditions

$$\Omega_1'^{+}(x) - \Omega_1'^{-}(x) = -\Omega_0'(x) - \overline{\Omega_0'(x)} - x\,\overline{\Omega_0''(x)} - \overline{\omega_0'(x)} \quad (3.22)$$

on $y = 0$ and have $O(1/z)$ as $|z| \to \infty$. This equation may be solved by means of the integral representations (3.16) used in the last section, but it is easier to note that $\Omega_0(z)$ is holomorphic in S^- and $z\,\overline{\Omega_0'(\bar{z})} + \overline{\omega_0(\bar{z})}$ is holomorphic in S^+, and hence we may satisfy the boundary conditions

(3.22) by putting†

$$\Omega_1(z)\begin{cases} = -z\,\overline{\Omega_0'(\bar{z})} - \overline{\omega_0(\bar{z})} & (z \in S^+) \\ = \Omega_0(z) & (z \in S^-). \end{cases} \tag{3.23}$$

Using equation (3.9) to define the corresponding $\omega_1(z)$, the total potentials describing the behaviour of the half-plane under the action of the point forces and moments are

$$\begin{aligned} \Omega(z) &= \Omega_0(z) - z\,\overline{\Omega_0'(\bar{z})} - \overline{\omega_0(\bar{z})} \\ \omega(z) &= \omega_0(z) + z\,\overline{\omega_0'(\bar{z})} - \overline{\Omega_0(\bar{z})} + z\,\overline{\Omega_0'(\bar{z})} + z^2\,\overline{\Omega_0''(\bar{z})}. \end{aligned} \tag{3.24}$$

Particular examples of this result are left to the reader. However it is interesting to note that for large $|z|$ from (2.58)

$$-\kappa\,\Omega_0(z) = \overline{\omega_0(\bar{z})} = \frac{\kappa P}{2\pi(1+\kappa)}\log z + O(1)$$

where P is the resultant applied force, and so from (3.24) we can conclude

$$-\Omega(z) = \overline{\omega(\bar{z})} = \frac{P}{2\pi}\log z + O(1)$$

which is in agreement with the results of Section 2.10 for a half-plane.

3.6 Displacement boundary value problem for a half-plane

Let us suppose the displacement $u + iv$ is specified at each point of the boundary $y = 0$ of the half-plane S^+. Then

$$u + iv = f(x) + ig(x) \quad \text{for all } x \text{ on } y = 0. \tag{3.25}$$

Now if we use the stress continuation described in Section 3.3 and resulting in equations (3.8), the stress and displacement fields may be described by (3.10) in terms of a single potential $\Omega(z)$ holomorphic in S^+ and S^-. Thus the boundary conditions (3.25) become

$$\kappa\,\Omega^+(x) + \Omega^-(x) = 2\mu\{f(x) + ig(x)\} \quad \text{for all } x \text{ on } y = 0 \tag{3.26}$$

provided $\lim_{y\to 0+} y\,\Omega'(z) = 0$ which must be confirmed after the solution is obtained. The values of $\Omega(z)$ in S^+ and in S^- are connected only through the boundary conditions (3.26) which hold on $y = 0$ and hence we can

† Note that as the stresses (3.21) have $O(1/x)$ for large $|x|$ we do not expect $\Omega_1(z)$ to satisfy (2.67) for large $|z|$.

define a new potential

$$\theta(z)\begin{cases} = \kappa\,\Omega(z) & (z \in S^+) \\ = -\Omega(z) & (z \in S^-) \end{cases} \tag{3.27}$$

so that the boundary conditions (3.26) become

$$\theta^+(x) - \theta^-(x) = 2\mu\{f(x) + \mathrm{i}g(x)\} \quad \text{for all } x \text{ on } y = 0. \tag{3.28}$$

It will be seen on comparing equations (3.27), (3.8) and (3.13) that $\theta(z)$ could have been defined directly from the displacement continuation outlined at the end of Section 3.3.

Now it has been shown in Section 1.6 that if $\theta(z) = O(1/z)$ as $|z| \to \infty$ the solution to (3.28) is of the form

$$\theta(z) = \frac{\mu}{\pi \mathrm{i}} \int_{-\infty}^{\infty} \frac{f(t) + \mathrm{i}g(t)}{t - z}\,\mathrm{d}t. \tag{3.29}$$

However it may be shown from (3.27) and (3.11) that if a finite resultant force acts on the half-space S^+ then $\theta(z)$ is proportional to $\log z$ as $|z| \to \infty$ and consequently the solution (3.29) can only apply when the displacements are such that the resultant force applied to S^+ is zero. The way around this difficulty is to specify the displacement gradients on $y = 0$, so that

$$\frac{\partial u}{\partial x} + \mathrm{i}\frac{\partial v}{\partial x} = f'(x) + \mathrm{i}g'(x) \quad \text{on } y = 0 \tag{3.30}$$

and from (3.27) and (3.10) the boundary conditions become

$$\theta'^+(x) - \theta'^-(x) = 2\mu\{f'(x) + \mathrm{i}g'(x)\} \quad \text{on } y = 0 \tag{3.31}$$

where $\theta'(z) = O(1/z)$ as $|z| \to \infty$.

Now applying the results of Section 1.6 we find

$$\theta'(z) = \frac{\mu}{\pi \mathrm{i}} \int_{-\infty}^{\infty} \frac{f'(t) + \mathrm{i}g'(t)}{t - z}\,\mathrm{d}t. \tag{3.32}$$

Again as in Section 3.4 we must assume $f'(t) + \mathrm{i}g'(t)$ is Hölder continuous for all t including the point at infinity. Particular examples are left to the reader. It is worth remarking that for both the displacement and mixed boundary value problems it proves convenient to satisfy the displacement condition (3.25) in the differentiated form (3.30) as $\Omega'(z)$ has a simple zero at infinity.

3.7 Mixed boundary value problems for a half-plane

Introduction

In the following three sections a variety of boundary value problems for the half-plane S^+ will be considered. In all cases the stress continuation method described in Section 3.3 will be used to represent the stress and displacement fields in terms of a single complex function that is sectionally holomorphic in the whole plane. This function is then found to satisfy the Hilbert problem for a set of line intervals on $y = 0$. For the purposes of illustration the problems may be regarded as the steady-state indentation of the half-plane by a set of rigid beams or punches. In this section we examine the case where the beams completely adhere to the half-plane. In Sections 3.8 and 3.9 we suppose the beams or punches either make a frictionless contact with the half-plane (Section 3.8) or are in a state of limiting friction under the action of a propelling force along the surface of the half-plane. Some alternative methods of solution of this group of problems are summarized in Section 3.11.

The remaining principal boundary value problem as described in Section 2.11 for the half-plane S^+ is the mixed boundary value problem in which the complex displacement $D = u + iv$ is specified over a region L of $y = 0$ and the stresses $\tau_{yy} - i\tau_{xy}$ are specified over the remainder L' of $y = 0$. As we are dealing with linear elasticity we may regard this problem as the sum of two problems. The first being a stress boundary value problem in which the given stresses are applied on L' and (say) zero stresses are applied on L. From the known solution to this problem we may calculate the displacements on L. The second problem is thus a mixed problem in which zero stresses are applied on L' and a displacement distribution is defined on L, these displacements being the difference between the specified displacements and those calculated from the first problem. Thus, without loss of generality, we may regard the mixed problem as having known displacements specified on L and zero stresses on the remainder L' of $y = 0$. A physical situation which corresponds to this problem is when a set of rigid punches of given profiles are brought into contact with the surface of the half-plane and allowed to indent the surface in such a way that the normal and tangential displacements of the surface are known at each point of the contact region L. This is the case for example if the body adheres to the punches over the entire region L on initial contact and during the subsequent indentation no slip occurs and the contact region L does not change.

Let us suppose L is the union of a finite set of line segments $L_k = (a_k, b_k)$, $k = 1, 2, \ldots, n$, where the ends of the segments are encountered in the order $a_1, b_1, a_2, b_2, \ldots, a_n, b_n$ when moving in the positive x-direction. The boundary conditions are then

$$\begin{aligned} D &= f_k(x) - ic_k \quad (x \in L_k, \quad k = 1, \ldots, n) \\ \tau_{yy} - i\tau_{xy} &= 0 \qquad\qquad\quad (x \in L') \end{aligned} \tag{3.33}$$

where $\mathrm{Im}\{f_k(x)\}$ may be interpreted as the profile of the kth (rigid) punch and c_k the relative depth of penetration if we choose $f_k(x)$ so that $\max_{x \in L_k} \text{Im}\{f_k(x)\} = 0$.

If we use the stress continuation described in Section 3.3 then, from (3.10) and (3.12), the boundary conditions become

$$\begin{aligned} \kappa\, \Omega^+(x) + \Omega^-(x) &= 2\mu(f_k(x) + \mathrm{i}c_k) \quad (x \in L_k, \quad k = 1, \ldots, n) \\ \Omega'^+(x) - \Omega'^-(x) &= 0 \quad (x \in L'). \end{aligned} \tag{3.34}$$

Hence the complex potential $\Omega'(z)$ is holomorphic in S^+ and S^- and continuous across region L' of $y = 0$, consequently $\Omega'(z)$ is sectionally holomorphic in the whole plane cut along the segments L_k ($k = 1, \ldots, n$) across which it satisfies the boundary condition (3.34). Also assuming the total force $X + \mathrm{i}Y$ applied to the half-space is finite we can conclude from (3.11)

$$\Omega'(z) = -\frac{X + \mathrm{i}Y}{2\pi z} + O\left(\frac{1}{z^2}\right) \tag{3.35}$$

for large $|z|$.

As in the previous section since $\Omega'(z)$ has a simple zero at infinity it is more convenient to solve the boundary condition (3.34) in differential form giving

$$\begin{aligned} \kappa\, \Omega'^+(x) + \Omega'^-(x) &= 2\mu f_k'(x) \quad (x \in L_k, \quad k = 1, \ldots, n) \\ &= 2\mu f'(x) \quad (x \in L). \end{aligned} \tag{3.36}$$

Now this relation between the boundary values of a holomorphic function has been considered in Section 1.6 and is referred to as the Hilbert problem. From (1.34) the most general function $\Omega'(z)$ satisfying (3.36) has the form

$$\Omega'(z) = \frac{\mu\, X(z)}{\kappa\pi \mathrm{i}} \int_L \frac{f'(t)\, \mathrm{d}t}{X^+(t)(t - z)} + X(z)P(z) \tag{3.37}$$

where $X(z)$ is the basic Plemelj function (1.35)

$$X(z) = \prod_{k=1}^{n} (z - a_k)^{-\gamma}(z - b_k)^{\gamma - 1} \tag{3.38}$$

and

$$2\pi \mathrm{i}\gamma = \log(-1/\kappa) = -\log|\kappa| + \mathrm{i}\pi$$

so that

$$\gamma = \tfrac{1}{2} + \mathrm{i}\beta, \qquad \beta = \frac{1}{2\pi} \log|\kappa| \tag{3.39}$$

and $P(z)$ is an arbitrary polynomial. As in Section 1.6 we consider the branch of $X(z)$ which is such that

$$z^n X(z) \to 1 \quad \text{as } |z| \to \infty.$$

Since

$$\Omega'(z) = O\left(\frac{1}{z}\right) \quad \text{as } |z| \to \infty$$

we see that $P(z)$ is at most a polynomial of degree $n - 1$

$$P(z) = D_0 + D_1 z + \cdots + D_{n-1} z^{n-1} \tag{3.40}$$

and also from (3.35)

$$D_{n-1} = -\frac{X + \mathrm{i}Y}{2\pi}. \tag{3.41}$$

As the remaining $(n - 1)$ coefficients of $P(z)$ in (3.40) are not determined by the equations (3.35) and (3.36), it is apparent that additional physical assumptions are required before the problem is solved completely.

Let us suppose first of all that the resultant forces applied to each punch are known. Then if $X_k + \mathrm{i}\,Y_k$ is the known resultant force across L_k we find

$$\begin{aligned} X_k + \mathrm{i}\,Y_k &= -\int_{L_k} (\tau_{xy} + \mathrm{i}\tau_{yy})\,\mathrm{d}x \\ &= -\mathrm{i}\int_{L_k} (\Omega'^+(x) - \Omega'^-(x))\,\mathrm{d}x \end{aligned} \tag{3.42}$$

for $k = 1, 2, \ldots, n$. On substituting from (3.37) these equations yield n equations for the determination of n constants D_j. Clearly one of these equations will be redundant as (3.41) ensures the overall equilibrium of the system with

$$X + \mathrm{i}Y = \sum_{k=1}^{n} X_k + \mathrm{i}\,Y_k.$$

However this system of n equations completely determines the solution (3.37).

Alternatively we may suppose that the relative depths of penetration of the punches have been specified so that the constants c_k in (3.33) are known to within an overall rigid-body displacement. Thus the coefficients of $P(z)$ must be chosen so that the continuous displacements on $y = 0$ satisfy (3.33). Now

$$2\mu\frac{\partial D}{\partial x} = \kappa\,\Omega'^+(x) + \Omega'^-(x) \quad \text{for all } x \text{ on } y = 0 \tag{3.43}$$

and on integrating with respect to x an expression for the complex displacement D on $y = 0$ is obtained. On integration along the intervals L_k, equations (3.34) are satisfied to within constants independently of $P(z)$. We must therefore integrate over the regions between the punches. If we do this we find

$$\int_{b_k}^{a_{k+1}} \frac{\partial D}{\partial x}\,\mathrm{d}x = D(a_{k+1}) - D(b_k)$$
$$= f_{k+1}(a_{k+1}) + \mathrm{i}c_{k+1} - f(b_k) - \mathrm{i}c_k \qquad (3.44)$$

for $k = 1, 2, \ldots, n-1$, where over the interval (b_k, a_{k+1})

$$\frac{\partial D}{\partial x} = \frac{(1+\kappa)}{2\mu}\,\Omega'^{+}(x).$$

Equations (3.44) are $(n-1)$ equations for the determination of the n constants D_j. The remaining equation is found by noting that the resultant force $X + \mathrm{i}Y$ over all punches will be known and hence D_{n-1} is given by (3.41). Again we see the solution (3.37) to the problem is completely determined by these conditions.

In deriving the solutions above we have assumed that the functions $f_k(x)$ are known completely. However it is quite possible when indenting with known forces and moments applied to each punch that $f_k(x)$ may be undetermined to within a rigid-body rotation $\mathrm{i}\varepsilon_k x$ for $k = 1, 2, \ldots, n$. The relationship between the applied moments and the rotations may be found by evaluating the moments across the contact regions L_k from the solution (3.37). Further details of such calculations may be examined in Muskhelishvili,[43] Chapter 19 and an example is given in the second of the following illustrations.

1. *A flat-ended rough punch*

We first examine the case of normal indentation by a single flat-ended punch which makes contact with S^+ over the region $|x| \leqslant a$. Then

$$f'(x) = 0$$

and from (3.37) and (3.41) we find

$$\Omega'(z) = -\frac{X + \mathrm{i}Y}{2\pi}\,X(z), \qquad (3.45)$$

where

$$X(z) = (z+a)^{-\frac{1}{2}-\mathrm{i}\beta}(z-a)^{-\frac{1}{2}+\mathrm{i}\beta}. \qquad (3.46)$$

The stress under the punch is simply found to be

$$\tau_{yy} - \mathrm{i}\tau_{xy} = \Omega'^{+}(x) - \Omega'^{-}(x) = (1+\kappa)\,\Omega'^{+}(x)$$

where if we put $z - a = R_1 e^{i\theta_1}$, $z + a = R_2 e^{i\theta_2}$

$$\begin{aligned} X(z) &= (R_1R_2)^{-1/2}\exp\{i\beta\log(R_1/R_2) + \beta(\theta_2 - \theta_1) - i(\theta_2 + \theta_1)/2\} \\ &= -ie^{-\pi\beta}(R_1R_2)^{-1/2}\exp\{i\beta\log(R_1/R_2)\} \quad \text{on } y = 0+, \quad |x| \leqslant a. \end{aligned} \tag{3.47}$$

Now using (3.39) we find on $y = 0$, $|x| \leqslant a$

$$\tau_{yy} - i\tau_{xy} = \frac{i(X + iY)}{2\pi\sqrt{\kappa}} \cdot \frac{(1 + \kappa)}{(a^2 - x^2)^{1/2}} e^{i\beta\log|(a-x)/(a+x)|}. \tag{3.48}$$

This expression is worth examining in some detail as it exposes an anomaly in the linear theory of elasticity. We see that in addition to the term $(a^2 - x^2)^{-1/2}$ which diverges at the corners $x = \pm a$ of the punch, terms of the form $\cos(\beta\log|(a - x)/(a + x)|)$ and $\sin(\beta\log|(a - x)/(a + x)|)$ are present which alternate in sign with increasing frequency as $x \to \pm a$. Thus the character of the stress singularities is that as $x \to \pm a$ the stress changes in sign an infinite number of times as its magnitude diverges, indicating that large tensile as well as large compressive forces occur in the regions near the corners of the punch. This physically unreal result is directly dependent both on the linearizing assumptions of the present theory and also on the perfect bonding assumption used in the boundary conditions. Many† other situations have been discovered in which a similar breakdown occurs, all of them occurring at points where either the elastic body has a sharp corner or the boundary conditions change in type or change discontinuously.

It is possible to evaluate the size of the regions over which the stress oscillations occur. For both plane strain and generalized plane stress we have shown in (2.40) that κ satisfies the inequality $1 < \kappa < 3$. From (3.48) if $X = 0$ the normal stress first changes sign when

$$\beta\log\left|\frac{a - x}{a + x}\right| = \pm\tfrac{1}{2}\pi$$

and hence

$$x = \pm a\tanh(\pi^2/2\log\kappa). \tag{3.49}$$

The smallest value of $|x|$ is obtained when κ attains its maximum value, namely $\kappa = 3$, and so $x = \pm 0{\cdot}9997a$. Thus in the extreme case when $\kappa = 3$ the region of stress oscillation is approximately 0·03 per cent of the contact region and for actual materials will be considerably smaller.

It should be noted however that although this effect is limited in extent it

† See for example Karp and Karal,[27] Erdogan,[15] Williams,[73] England.[10]

occurs at the points at which we are most interested in the behaviour of the body, and prevents us from making any mathematically sound predictions about the nature of the deformation near these points.

2. *A flat-ended beam tilted by a couple*

A second problem illustrating the above theory is the case of a flat-ended beam which adheres to the half-plane S^+ and is then tilted by the application of a couple M. Let us suppose the beam is of width $2a$ and is tilted through a small angle ε. Then in (3.34)

$$f(x) = \mathrm{i}\varepsilon x \quad (|x| \leqslant a)$$

and from (3.37)

$$\Omega'(z) = \frac{\mu\varepsilon\, X(z)}{\kappa\pi} \int_{-a}^{a} \frac{\mathrm{d}t}{X^+(t)(t-z)} + D_0\, X(z) \tag{3.50}$$

where $X(z)$ has been defined in (3.46). If we suppose the resultant force applied to S^+ is zero (this case having been examined in the first illustration) then from (3.41) $D_0 = 0$. A method of evaluating a contour integral of the type in (3.50) has been discussed in Section 1.8 and applying the results of that section we find

$$\Omega'(z) = \frac{2\mu\varepsilon\mathrm{i}}{1+\kappa}\{1 - (z + 2\mathrm{i}a\beta)X(z)\}. \tag{3.51}$$

This now enables us to calculate the stress over the contact region and hence evaluate the relation between the angle of tilt ε and the applied couple M. Over the contact region

$$\begin{aligned} \tau_{yy} - \mathrm{i}\tau_{xy} &= \Omega'^{+}(x) - \Omega'^{-}(x) \\ &= -2\mu\mathrm{i}\varepsilon(x + 2\mathrm{i}a\beta)X^+(x) \end{aligned}$$

and hence

$$\begin{aligned} M &= -\int_{-a}^{a} x\tau_{yy}\, \mathrm{d}x \\ &= 2\mu\varepsilon\, \mathrm{Im}\left\{\int_{-a}^{a} x(x + 2\mathrm{i}a\beta)X^+(x)\, \mathrm{d}x\right\}. \end{aligned}$$

As in Section 1.8 we may evaluate this integral by replacing the integral over the interval $(-a, a)$ by a contour integral over a lacet C around $(-a, a)$, see Figure 1.6, and hence we find

$$M = \frac{2\mu\varepsilon}{1+\kappa}\, \mathrm{Im}\left\{\int_C z(z + 2\mathrm{i}a\beta)X(z)\, \mathrm{d}z\right\}.$$

We can now replace C by a contour of large radius about the origin, and for large $|z|$ the coefficient of $1/z$ in the integrand is found to be $(1 + 4\beta^2)a^2/2$. Thus

$$M = \frac{2\mu\pi\varepsilon(1 + 4\beta^2)a^2}{1 + \kappa}$$

and, for a given couple M, the angle of tilt ε is determined by

$$\varepsilon = \frac{(1 + \kappa)M}{2\pi\mu(1 + 4\beta^2)a^2}. \tag{3.52}$$

By combining this solution with that of the first illustration we may examine the effect of a rough flat-ended punch or beam on the half-plane under the action of a given force and moment.

An illustration of the calculations for a finite number of punches has been omitted since the theory has been given above and such problems merely reduce to more complicated contour integration and yield little further appreciation of the methods of solution. A two-punch problem is considered in Example 6.

3.8 Frictionless punch problems

Let us consider first the case of a rigid frictionless symmetric punch normally indenting the half-plane S^+ under the action of a resultant force Y. If the contact surface of the punch has the equation $y = f(x)$, where by assumption $f(x)$ is an even function and for completeness we take $f(0) = 0$, then the contact region is symmetrical and for simplicity we can suppose contact is maintained over the interval $|x| \leqslant a$. Thus the half-plane S^+ suffers the indentation

$$v = f(x) + C \quad \text{on } y = 0, \ |x| \leqslant a$$

where C is a real constant. As the punch is assumed frictionless the shear stress under the punch is zero, and without loss of generality we may assume the remainder of $y = 0$ is unstressed. Thus the boundary conditions for the problem are

$$\left.\begin{aligned} v &= f(x) + C \\ \tau_{xy} &= 0 \end{aligned}\right\} \quad (y = 0, \ |x| \leqslant a) \tag{3.53}$$

$$\tau_{yy} - i\tau_{xy} = 0 \qquad (y = 0, \ |x| > a). \tag{3.54}$$

If we now make use of the stress continuation derived in Section 3.3, on $y = 0+$ we find

$$\begin{aligned} \tau_{yy} - i\tau_{xy} &= \Omega'^+(x) - \Omega'^-(x) \\ 2\mu(u + iv) &= \kappa\,\Omega^+(x) + \Omega^-(x) \end{aligned} \tag{3.55}$$

where $\Omega'(z)$ is a holomorphic function defined in S^+ and S^-. Also for large $|z|$ it has been shown in (3.11) that

$$\Omega'(z) = -\frac{\mathrm{i}Y}{2\pi z} + O\left(\frac{1}{z^2}\right). \tag{3.56}$$

From the boundary condition (3.54) it is clear that $\Omega'(z)$ is holomorphic in the whole plane cut along the arc $y = 0$, $|x| \leqslant a$. The complex potential $\Omega(z)$ must now be chosen so that the conditions (3.53) are satisfied. From (3.55) we find

$$\tau_{xy} = -\frac{1}{2\mathrm{i}}\{\Omega'^+(x) - \overline{\Omega'^+(x)} - \Omega'^-(x) + \overline{\Omega'^-(x)}\}$$

and hence the condition $\tau_{xy} = 0$ on $y = 0+$, $|x| \leqslant a$ implies

$$\lim_{y\to 0+} \{\Omega'(z) + \overline{\Omega'(\bar{z})}\} = \lim_{y\to 0-} \{\Omega'(z) + \overline{\Omega'(\bar{z})}\}.$$

Thus the function $\Omega'(z) + \overline{\Omega'(\bar{z})}$ is holomorphic in the whole plane and from (3.56) has $O(1/z^2)$ as $|z| \to \infty$. Hence by Laurent's theorem

$$\Omega'(z) + \overline{\Omega'(\bar{z})} = 0. \tag{3.57}$$

The problem has now reduced to determining a function $\Omega(z)$ holomorphic in the whole plane cut along $y = 0$, $|x| \leqslant a$ which satisfies (3.57) and the first condition of (3.53).

Again in view of the relation (3.57) and the known behaviour of $\Omega'(z)$ at infinity (3.56) it will be simpler to satisfy the first condition of (3.53) in differentiated form giving

$$\Omega'^+(x) + \Omega'^-(x) = \frac{4\mathrm{i}\mu}{1+\kappa} f'(x) \quad (y = 0, |x| \leqslant a) \tag{3.58}$$

on noting from (3.57) that $\Omega'^+(x) = -\overline{\Omega'^-(x)}$.

Hence $\Omega'(z)$ satisfies a Hilbert problem of the form considered in Section 1.6 and from (1.34) the general solution of (3.58) is

$$\Omega'(z) = \frac{2\mu}{(1+\kappa)\pi\, X(z)} \int_{-a}^{a} \frac{X^+(t)f'(t)\,\mathrm{d}t}{t - z} + \frac{P(z)}{X(z)}, \tag{3.59}$$

where in this case

$$X(z) = (z^2 - a^2)^{1/2} \tag{3.60}$$

and we select the branch such that $z^{-1}X(z) \to 1$ as $|z| \to \infty$. Also $P(z)$ is an arbitrary polynomial, but in view of the known behaviour of $\Omega'(z)$ at infinity we see that $P(z)$ is at most a constant, and in fact from (3.56)

$$P(z) = -\frac{\mathrm{i}Y}{2\pi}. \tag{3.61}$$

We must now confirm that (3.57) holds. In order to calculate $\overline{\Omega'(\bar{z})}$ from (3.59) we must determine the values of $\overline{X(\bar{z})}$ and $\overline{X^+(t)}$ for the given branch. Now from (3.60) we see that

$$X^+(t) = \mathrm{i}(a^2 - t^2)^{1/2} \quad (|t| \leqslant a)$$

and hence

$$\overline{X^+(t)} = -X^+(t).$$

Similarly it can be shown that

$$\overline{X(\bar{z})} = X(z)$$

and since from (3.61) $P(z)$ is imaginary we can conclude that (3.57) is satisfied. Thus the solution to the problem of symmetric indentation of the half-plane S^+ by a frictionless punch having an end section $y = f(x)$ under a resultant force Y is

$$\Omega'(z) = \frac{2\mathrm{i}\mu}{(1+\kappa)\pi(z^2 - a^2)^{1/2}} \int_{-a}^{a} \frac{(a^2 - t^2)^{1/2} f'(t)\,\mathrm{d}t}{t - z} - \frac{\mathrm{i}Y}{2\pi(z^2 - a^2)^{1/2}}. \tag{3.62}$$

Having achieved this solution it must be confirmed that the material is under compression (i.e. $\tau_{yy} < 0$) over the whole contact region $|x| \leqslant a$ for the solution to be physically reasonable. This leads to a general inequality between $f(x)$ and the normal force Y which we shall not calculate. It seems physically obvious that this will be the case if $f''(x) \leqslant 0$. An alternative method of solution is given in Examples 7 and 8.

If we consider the indentation by a symmetric punch as shown in Figure 3.2 it will be seen that two distinct situations can occur. It is clear in Figure 3.2(a) that for small values of Y only a small portion of the end face of

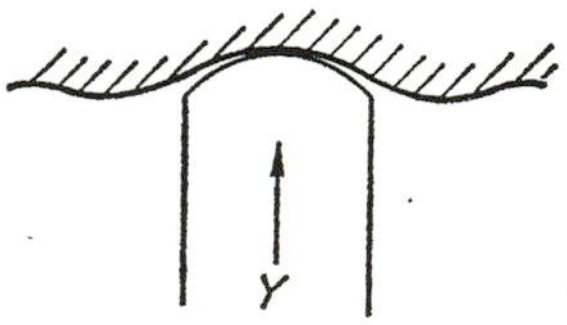

Incomplete penetration

Figure 3.2a

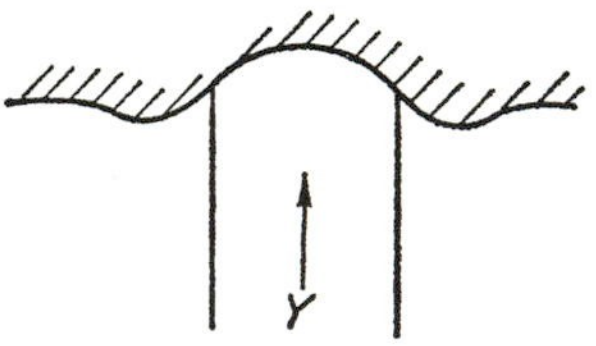

Complete penetration

Figure 3.2b

the punch will come into contact with the surface and it will be seen that the length of the region of contact depends on the applied force Y. The actual length of the contact region may be determined by ensuring that the normal stress is finite at the points where the surface leaves the punch smoothly. We shall refer to this situation as incomplete penetration, and a representative example is considered in the second illustration.† If as in Figure 3.2(*b*) the punch has corners then on increasing Y sufficiently the whole end-face of the punch will come into contact with the half-plane. We shall refer to this as complete penetration and it is clear from (3.62) that unbounded stresses will occur at these corners. We now consider two illustrations of these results.

1. *Smooth flat-ended punch*

If we suppose the punch (of width $2a$) is flat-ended then $f'(x) = 0$ and from (3.62) for normal indentation under the resultant force Y

$$\Omega'(z) = -\frac{\mathrm{i}Y}{2\pi(z^2 - a^2)^{1/2}}. \tag{3.63}$$

Thus

$$\Omega(z) = -\frac{\mathrm{i}Y}{2\pi}\log(z + (z^2 - a^2)^{1/2}) \tag{3.64}$$

to within an arbitrary constant which corresponds to a rigid-body displacement. From (3.64) and the general relations (3.10) the complete stress and displacement fields in S^+ may be calculated. It is of interest to evaluate the stress under the punch and from (3.55) we find

$$\tau_{yy} = -\frac{Y}{\pi(a^2 - x^2)^{1/2}} \quad (|x| \leqslant a). \tag{3.65}$$

As might be expected since this is a case of complete penetration there are stress singularities at the corners of the punch. However these are simple root-type singularities and not the essential singularities that were encountered in case of adhesion by a flat-ended punch, see Section 3.7 equation (3.48). Again it should be noted that the assumptions of linear elasticity are violated at these points.

In a similar manner the displacements on $y = 0$ may be simply evaluated and we record them here for completeness. Under the punch

$$u = \frac{Y(\kappa - 1)}{4\pi\mu}\theta, \qquad v = -\frac{Y(1 + \kappa)}{4\pi\mu}\log a \tag{3.66}$$

where $a \sin\theta = (a^2 - x^2)^{1/2}$ and θ takes values in the range $0 \leqslant \theta \leqslant \pi/2$,

† For the general case see Muskhelishvili[43], section 116.

$\pi/2 \leqslant \theta \leqslant \pi$ for $x \geqslant 0$, $x \leqslant 0$ respectively. On the free boundary $|x| \geqslant a$

$$u = Y(\kappa - 1)/4\mu \quad (x < -a), \qquad u = 0 \quad (x > -a),$$
$$v = -\frac{Y(\kappa + 1)}{4\pi\mu}\log(|x| + (x^2 - a^2)^{1/2}). \tag{3.67}$$

Equations (3.66) and (3.67) hold to within a rigid-body displacement. Note that a force-displacement relation for a punch acting on a half-plane cannot be found since the punch produces an infinite displacement of the surface $y = 0$. Nevertheless the stress field near the punch agrees with experimental observations. These comments apply to all two-dimensional punch problems.

2. *Smooth parabolic punch*

If we suppose that a symmetric smooth punch of width $2l$ has a radius of curvature R at $x = 0$ then it is consistent with the linear theory of elasticity to suppose R is large and to approximate the end section of the punch by the parabolic curve $y = -x^2/2R$ $(|x| \leqslant l)$. Let us suppose on indentation under the force Y the length of the region of contact is $2a$ $(\leqslant 2l)$. Then from (3.62)

$$\Omega'(z) = -\frac{2i\mu}{R(1 + \kappa)\pi(z^2 - a^2)^{1/2}}\int_{-a}^{a}\frac{(a^2 - t^2)^{1/2}t\,dt}{t - z} - \frac{iY}{2\pi(z^2 - a^2)^{1/2}}.$$

Now the contour integral

$$\int_{-a}^{a}\frac{t(a^2 - t^2)^{1/2}\,dt}{t - z}$$

may be evaluated by converting it into a contour integral of the form

$$\int_C \frac{\zeta(\zeta^2 - a^2)^{1/2}\,d\zeta}{\zeta - z}$$

where C is a lacet circumscribing the interval $(-a, a)$ in the manner shown in Figure 1.6. By use of the theory of residues we find

$$\Omega'(z) = -\frac{2i\mu}{(1 + \kappa)R}\left(z - \frac{2z^2 - a^2}{2(z^2 - a^2)^{1/2}}\right) - \frac{iY}{2\pi(z^2 - a^2)^{1/2}}. \tag{3.68}$$

We may now determine the length $2a$ of the region of contact by stipulating that the stresses should be bounded at the ends of the contact region. The stress under the punch is

$$\tau_{yy} - i\tau_{xy} = \Omega'^{+}(x) - \Omega'^{-}(x)$$
$$= \frac{1}{\pi(a^2 - x^2)^{1/2}}\left\{\frac{2\pi\mu(2x^2 - a^2)}{R(1 + \kappa)} - Y\right\} \tag{3.69}$$

and hence for boundedness $a^2 = (1 + \kappa)YR/2\pi\mu$. Thus the normal stress $\tau_{yy} = 0$ at $x = \pm a$ and is continuous on $y = 0$. This relation clearly implies that for incomplete penetration

$$Y \leqslant 2\pi\mu l^2/(1 + \kappa)R \tag{3.70}$$

and if Y exceeds this value the whole end face of the punch comes into contact with S^+ and the stress under the punch is given by (3.69) on replacing a by l. It may be confirmed in this case that there are root-type stress singularities at $x = \pm l$.

We have restricted consideration in this section to symmetric indentation by a single smooth punch but it will be seen that this work may be first generalized to an unsymmetrical indentation by a single punch merely by taking the contact region to be a general interval (a, b). In this case if one or both ends of the contact region are in smooth contact with the half-plane their positions may be determined by requiring that the normal stress be bounded at these points. A calculation of this nature is given in Section 3.9, see also Example 9. The generalization to a finite number of punches follows directly from the considerations of Section 3.7. Further details may be found in Muskhelishvili,[43] Section 115, and Galin.[17] Problems of an infinite row of frictionless punches acting on the half-space have been considered by Schtaerman,[51] Chapter 2 and also by England and Green[12] who use a particular integral representation.

3.9 A sliding punch with friction

The problem of a set of punches sliding under conditions of limiting friction along the surface of a half-plane has been considered by Muskhelishvili,[43] Section 117, and using a different method by Galin,[17] Chapter I, Section 7. They suppose that over the contact regions the shear stress $s(x)$ is related to the pressure $p(x)$ under the kth punch by

$$s(x) = p(x) \tan \lambda_k,$$

where λ_k is the angle of limiting friction for the kth punch and is a constant under the punch. It was assumed that the punches move with a uniform velocity which is small compared with the speed of sound in the medium so that the static equations of elasticity are approximately true. For clarity we examine the case of indentation by a single punch in contact with the half-plane S^+ over the interval $(-a, b)$ under the action of a resultant normal force Y, and propelled from left to right by a tangential force $X = Y \tan \lambda$ see Figure 3.3. Then if $y = f(x)$ is the equation of the end-

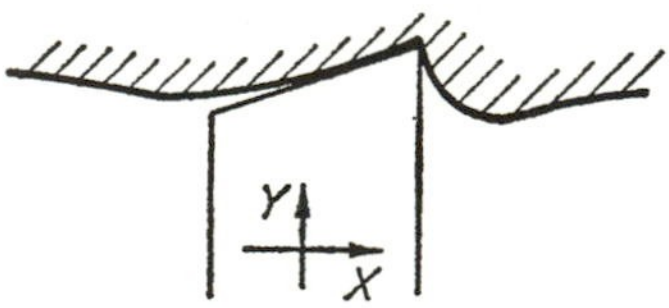

Figure 3.3a

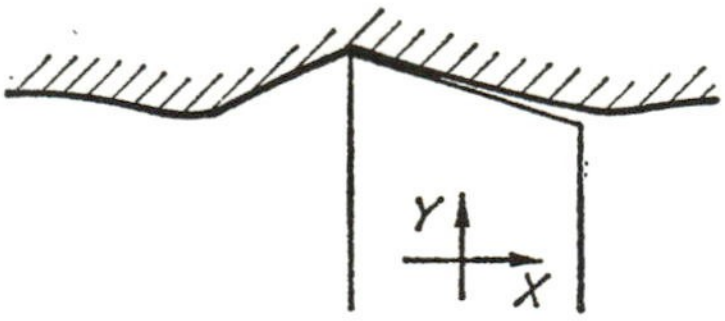

Figure 3.3b

face of the punch the boundary conditions of the problem are

$$\begin{aligned} v &= f(x) + \text{constant} \\ \tau_{xy} &= \tau_{yy}\tan\lambda \qquad\qquad y = 0+, \quad x \in (-a, b) \\ \tau_{yy} - \mathrm{i}\tau_{xy} &= 0 \qquad\qquad y = 0+, \quad x \notin (-a, b). \end{aligned} \tag{3.71}$$

We note that the second equation of (3.71) only holds provided $\tau_{yy} < 0$ an inequality which must be checked when the solution is obtained.

As in the previous sections we use the stress continuation of Section 3.3 to represent the stress and displacement fields in terms of a single function $\Omega(z)$ holomorphic in S^+ and S^-. From (3.12) we find

$$\begin{aligned} \tau_{yy} - \mathrm{i}\tau_{xy} &= \Omega'^+(x) - \Omega'^-(x) \\ 2\mu\frac{\partial D}{\partial x} &= \kappa\,\Omega'^+(x) + \Omega'^-(x). \end{aligned} \tag{3.72}$$

Thus from the last condition of (3.71), $\Omega'(z)$ is holomorphic in the whole plane cut along the contact region $y = 0$, $-a \leqslant x \leqslant b$. Now from (3.72) the stress τ_{yy} under the punch may be represented in the form

$$\tau_{yy} = \tfrac{1}{2}\{\Omega'^+(x) + \overline{\Omega'^+(x)} - \Omega'^-(x) - \overline{\Omega'^-(x)}\}$$

and similarly for τ_{xy}. Hence the boundary condition $\tau_{xy} = \tau_{yy}\tan\lambda$ implies

$$\begin{aligned} &\mathrm{i}\{\Omega'^+(x) - \overline{\Omega'^+(x)} - \Omega'^-(x) + \overline{\Omega'^-(x)}\} \\ &\qquad = \{\Omega'^+(x) + \overline{\Omega'^+(x)} - \Omega'^-(x) - \overline{\Omega'^-(x)}\}\tan\lambda \end{aligned}$$

which may be written in the form

$$\lim_{y\to 0+} \Omega'(z) + \mathrm{e}^{-2\mathrm{i}\lambda}\,\overline{\Omega'(\bar{z})} = \lim_{y\to 0-} \Omega'(z) + \mathrm{e}^{-2\mathrm{i}\lambda}\,\overline{\Omega'(\bar{z})}$$

as $(\mathrm{i} + \tan\lambda)/(\mathrm{i} - \tan\lambda) = \mathrm{e}^{-2\mathrm{i}\lambda}$.

Thus the function $\Omega'(z) + \mathrm{e}^{-2\mathrm{i}\lambda}\,\overline{\Omega'(\bar{z})}$ is holomorphic in the whole plane. As we have assumed a finite force acts on the half-plane S^+ and the stress and rotation are zero at infinity then $\Omega'(z) = O(z^{-1})$ as $|z| \to \infty$ and hence we may conclude

$$\Omega'(z) + \mathrm{e}^{-2\mathrm{i}\lambda}\,\overline{\Omega'(\bar{z})} = 0. \tag{3.73}$$

Thus the problem reduces to determining a sectionally holomorphic potential $\Omega(z)$ satisfying (3.73) and the displacement condition of (3.71). Again since we know $\Omega'(z) = O(z^{-1})$ as $|z| \to \infty$ this condition is most easily satisfied in differentiated form. From (3.72) and (3.73) we find

$$4i\mu \frac{\partial v}{\partial x} = (\kappa + e^{2i\lambda})\, \Omega'^{+}(x) + (1 + \kappa e^{2i\lambda})\, \Omega'^{-}(x)$$

and the boundary condition (3.71) becomes

$$\Omega'^{+}(x) + \alpha\, \Omega'^{-}(x) = \frac{4i\mu}{\kappa + e^{2i\lambda}} f'(x) \quad (x \in (-a, b)) \tag{3.74}$$

where

$$\alpha = \frac{1 + \kappa e^{2i\lambda}}{\kappa + e^{2i\lambda}}. \tag{3.75}$$

This will be recognized as a Hilbert problem of the sort considered in Section 1.6 and from (1.34) has the solution

$$\Omega'(z) = \frac{4i\mu\, X(z)}{(\kappa + e^{2i\lambda})2\pi i} \int_{-a}^{b} \frac{f'(t)\, dt}{X^{+}(t)(t - z)} + P(z)X(z) \tag{3.76}$$

where
$$X(z) = (z + a)^{-\gamma}(z - b)^{\gamma - 1} \tag{3.77}$$

$$2\pi i\gamma = \log(-\alpha) \quad (0 \leqslant \mathrm{Re}(\gamma) < 1) \tag{3.78}$$

and we select the branch of $X(z)$ such that $z\, X(z) \to 1$ as $|z| \to \infty$. Also since the resultant force applied to S^{+} is $X + iY = Y(\tan \lambda + i)$, from (3.11) $\Omega'(z)$ has the form

$$\Omega'(z) = -\frac{Y(\tan \lambda + i)}{2\pi z} + O\left(\frac{1}{z^2}\right)$$

for large $|z|$ and hence

$$P(z) = -\frac{Y(\tan \lambda + i)}{2\pi}. \tag{3.79}$$

It is of interest at this point to note that

$$|\alpha| = \left|\frac{1 + \kappa e^{2i\lambda}}{\kappa + e^{2i\lambda}}\right| = 1$$

and hence γ is real and is given by

$$\gamma = \frac{1}{2\pi} \arg(-\alpha), \quad 0 \leqslant \gamma < 1. \tag{3.80}$$

We must now check that the solution defined by (3.76) satisfies the condition (3.73). This may be confirmed on noting that $\overline{X(\bar{z})} = X(z)$ and that $\overline{X^{+}(t)} = -X^{+}(t)/\alpha$.

The problem is now solved in principle and particular cases may be investigated by assuming different shapes of punch and regions of contact. It should be noticed that in the absence of friction, $\lambda = 0$, the solution reduces to that obtained in the previous section. We shall now examine two cases of incomplete penetration where the region of contact is unknown and has to be determined by assuming the stresses are bounded at the edges of the contact region.

1. *Wedge-shaped punch*

We consider the motion to the right of the wedge-shaped punch shown in Figure 3.3(*a*) where the end section of the punch is $f(x) = \varepsilon x$ and we take the origin so that the contact region is $-l \leqslant x \leqslant l$. Then from (3.77) $X(z) = (z + l)^{-\gamma}(z - l)^{\gamma-1}$ and from (3.76) $\Omega'(z)$ depends on the integral

$$\int_{-l}^{l} \frac{\mathrm{d}t}{X^+(t)(t - z)}$$

which has already been evaluated in Section 1.8. Hence

$$\Omega'(z) = \frac{4\mathrm{i}\mu\varepsilon}{(\kappa + \mathrm{e}^{2\mathrm{i}\lambda})(1 + \alpha)} [1 - \{z + (2\gamma - 1)l\}X(z)] - \frac{Y(\tan \lambda + \mathrm{i})}{2\pi} X(z). \quad (3.81)$$

Under the punch the stress distribution is

$$\begin{aligned} \tau_{yy} - \mathrm{i}\tau_{xy} &= \Omega'^+(x) - \Omega'^-(x) \\ &= - \left[\frac{4\mathrm{i}\mu\varepsilon\{x + (2\gamma - 1)l\}}{(1 + \kappa)} + \frac{\mathrm{i}Y}{\pi}\right] \frac{\{X^+(x) - X^-(x)\}}{(1 + \mathrm{e}^{2\mathrm{i}\lambda})} \end{aligned}$$

and on using the bipolar coordinates $z - l = R_1\mathrm{e}^{\mathrm{i}\theta_1}$, $z + l = R_2\mathrm{e}^{\mathrm{i}\theta_2}$ it will be seen that $X^+(x) = |x + l|^{-\gamma}|x - l|^{\gamma-1}\mathrm{e}^{\mathrm{i}\pi(\gamma-1)}$ for $|x| \leqslant l$. Also since $\tau_{xy} = \tau_{yy} \tan \lambda$ the normal stress under the punch is found to be

$$\tau_{yy} = -\frac{\sin \pi\gamma}{1 + \kappa}\left[4\mu\varepsilon\{x + (2\gamma - 1)l\} + \frac{Y(1 + \kappa)}{\pi}\right] \frac{1}{|x + l|^{\gamma}|x - l|^{1-\gamma}} \quad (3.82)$$

which agrees in form with Muskhelishvili,[43] Section 117.

Now this solution as it stands applies to the case of complete indentation by a wedge-shaped punch of width $2l$ where both corners of the end-face touch the half-space, there being compressive stresses under the punch provided $Y(1 + \kappa) > 8\pi\mu\varepsilon(1 - \gamma)l$. However if Y is not sufficiently large to satisfy this inequality a state of incomplete penetration will result as illustrated in Figure 3.3(*a*). In this case the length of the contact region will depend on Y and is determined from the condition that the stress is bounded at the point $x = -l$ where the punch and half-plane meet smoothly. For a bounded stress at $x = -l$, from (3.82)

$$l = \frac{Y(1+\kappa)}{8\pi\mu\varepsilon(1-\gamma)} \tag{3.83}$$

and hence

$$\tau_{yy} = -\frac{4\mu\varepsilon\sin(\pi\gamma)}{(1+\kappa)}\left|\frac{x+l}{x-l}\right|^{1-\gamma}. \tag{3.84}$$

It is of interest to determine how the length of the contact region changes as the coefficient of friction λ increases. Now from Figure 3.4

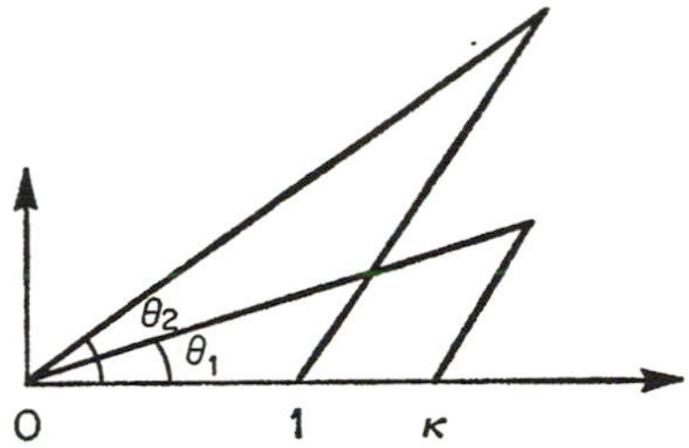

Figure 3.4

$$\begin{aligned}\gamma &= \frac{1}{2\pi}\arg\left(-\frac{1+\kappa e^{2i\lambda}}{\kappa+e^{2i\lambda}}\right)\\ &= \frac{1}{2\pi}(\pi+\theta_2-\theta_1) \quad (0 \leqslant \gamma < 1)\end{aligned} \tag{3.85}$$

where

$$\theta_2 = \arg(1+\kappa e^{2i\lambda}), \qquad \theta_1 = \arg(\kappa + e^{2i\lambda}).$$

As $\kappa > 1$, $(\theta_2 - \theta_1)$ is an increasing function of λ for $0 \leqslant \lambda < \pi/2$ in the range $0 \leqslant (\theta_2 - \theta_1) < \pi$. Hence as λ increases from 0 to $\pi/2$, γ increases from $\frac{1}{2}$ to 1, and from (3.83) for a fixed normal force Y the half-length of the contact region increases from $Y(1+\kappa)/4\pi\mu\varepsilon$ to infinity. It may also be confirmed from (3.84) that $\tau_{yy} < 0$ under the punch and τ_{xy} is negative corresponding to motion to the right.

If we rotate the punch so that on moving to the right the slant-edge meets the half-plane Figure 3.3(*b*), the same calculation may be performed and the half-length of the contact region found to be

$$l = \frac{Y(1+\kappa)}{8\pi\mu\varepsilon\gamma}.$$

In this case as the coefficient of friction λ increases from 0 to $\pi/2$, l decreases from $Y(1+\kappa)/4\pi\mu\epsilon$ in the frictionless case to $Y(1+\kappa)/8\pi\mu\varepsilon$ for $\lambda = \pi/2$, for a fixed normal force Y. Again these results are not in agreement with one's immediate physical intuition due to the way the surface curls under the applied shear stress.

2. *Parabolic punch*

As a second example we consider indentation by a parabolic punch having an end-face $f(x) = -x^2/2R$ and suppose the contact region is the interval $(-a, b)$ when the punch is moving to the right—see Figure 3.2. In this case to evaluate $\Omega'(z)$ we require

$$\int_{-a}^{b} \frac{t\,\mathrm{d}t}{X^+(t)(t-z)} = \frac{1}{1+\alpha}\int_C \frac{\zeta\,\mathrm{d}\zeta}{X(\zeta)(\zeta - z)}$$

where C is the lacet around the interval $(-a, b)$. This integral may be evaluated by the method used in Section 1.8 and hence we find

$$\begin{aligned}\Omega'(z) = &-\frac{4\mathrm{i}\mu}{R(\kappa + e^{2\mathrm{i}\lambda})(1+\alpha)}\\ &\times\left[z - \left\{z^2 + z(\gamma a + b(\gamma - 1)) + \frac{\gamma(\gamma-1)}{2}(a+b)^2\right\}X(z)\right]\\ &-\frac{Y(\tan\lambda + \mathrm{i})}{2\pi}X(z).\end{aligned} \tag{3.86}$$

Now on $y = 0$

$$\tau_{yy} - \mathrm{i}\tau_{xy} = \Omega'^+(x) - \Omega'^-(x)$$

and is bounded at the ends $x = -a$, $x = b$ of the contact region if

$$x^2 + x\{\gamma a + (\gamma - 1)b\} + \frac{\gamma(\gamma-1)}{2}(a+b)^2 = \frac{YR(\mathrm{i} + \tan\lambda)(\kappa + \mathrm{e}^{2\mathrm{i}\lambda})(1+\alpha)}{8\mathrm{i}\pi\mu}$$

when $x = -a$, $x = b$. Hence

$$\frac{a}{b} = \frac{\gamma}{1-\gamma}, \qquad ab = \frac{(1+\kappa)YR}{2\pi\mu} \tag{3.87}$$

and corresponding to these values we find

$$\tau_{yy} = -\frac{4\mu\sin(\pi\gamma)}{R(1+\kappa)}|x+a|^{1-\gamma}|b-x|^{\gamma} \tag{3.88}$$

which is always negative. It will be seen that as λ increases from 0 to $\pi/2$, γ increases from $\frac{1}{2}$ to 1 and hence b is a decreasing and a an increasing function of λ. Thus when the punch moves to the right for a fixed normal force Y, as the coefficient of friction increases the contact region moves to the left. In particular when $\lambda = \pi/2$ the contact region occupies the whole left half of the punch. Also when $\lambda = 0$ we find $a^2 = b^2 = (1+\kappa)YR/2\pi\mu$ which agrees with the results derived in Section 3.8.

It is worth noting that for problems of incomplete penetration, where it is known that stress singularities do not occur, the solution to the Hilbert problem (3.74) may be expressed (3.76) in terms of a Plemelj function which has zeros at the appropriate end points. These Plemelj functions have been given in Section 1.6. The location of the region of contact may be determined from the known behaviour of $\Omega'(z)$ at infinity.

In the last three sections it has been assumed that the indenting punch is rigid, however the methods may be generalized to deal with the Hertz problem of the contact of two dissimilar elastic media.†

3.10 Crack problems

The problems of an elastic body containing cracks have attracted much attention in view of their importance in a study of fracture, and various theories have been proposed to explain the generation of cracks in bodies and the conditions under which they will propagate.‡ In this section we are concerned with an infinite elastic plate containing a number of collinear line cracks or slits situated along the segments $L_r = (a_r, b_r)$, $r = 1, 2, \ldots, n$ of the real axis. The upper and lower surfaces of a crack are unconnected and regarded as external surfaces of the body and must satisfy the necessary physical condition that they do not cross.

If we suppose the plate containing unstressed collinear cracks is deformed by known stresses at infinity it will be seen that the problem is equivalent to the sum of two problems. The first being the deformation of an unbroken plate under the known stresses at infinity which gives rise to stress distributions $\tau_{yy} - i\tau_{xy} = f(x) + ig(x)$ across the segments L_r. The second being the deformation of the plate containing the cracks

† Muskhelishvili,[43] Section 119, Milne-Thomson,[41] Section 4.50.

‡ See for example Griffith,[22] Irwin,[25] Barenblatt,[1] and the recent series of articles edited by Liebowitz.[36]

with zero stresses at infinity, the cracks being opened by the stress distributions $\tau_{yy} - i\tau_{xy} = -\{f(x) + ig(x)\}$ acting on the upper and lower surfaces respectively. It will be seen that the resultant force over each crack is zero. As the solution to the first problem may be found by the methods of Section 3.1, it is sufficient to determine a solution to the second problem. We consider the more general situation in which the n collinear cracks are opened by the stress distributions $\tau_{yy} - i\tau_{xy} = -\{f_1(x) + ig_1(x)\}$, $\tau_{yy} - i\tau_{xy} = -\{f_2(x) + ig_2(x)\}$ acting on the upper and lower surfaces of the cracks respectively.

Starting from the basic equations (2.44) namely

$$\begin{aligned} 2\mu D &= \kappa\, \Omega(z) - z\,\overline{\Omega'(z)} - \overline{\omega(z)} \\ \tau_{yy} - i\tau_{xy} &= \Omega'(z) + \overline{\Omega'(z)} + z\,\overline{\Omega''(z)} + \overline{\omega'(z)} \end{aligned} \tag{3.89}$$

the complex potentials must be chosen so that the displacements and stresses are continuous across $y = 0$ outside of the cracks. Examining the displacements on $y = 0\pm$ we find

$$\kappa\, \Omega^+(x) - x\,\overline{\Omega'^+(x)} - \overline{\omega^+(x)} = \kappa\, \Omega^-(x) - x\,\overline{\Omega'^-(x)} - \overline{\omega^-(x)}$$

but this equation may be written in the form

$$\lim_{y\to 0+} \kappa\, \Omega(z) + z\,\overline{\Omega'(\bar z)} + \overline{\omega(\bar z)} = \lim_{y\to 0-} \kappa\, \Omega(z) + z\,\overline{\Omega'(\bar z)} + \overline{\omega(\bar z)}.$$

Thus the function $\kappa\, \Omega(z) + z\,\overline{\Omega'(\bar z)} + \overline{\omega(\bar z)}$ is continuous across $y = 0$ outside of the cracks. Therefore we may put

$$\kappa\, \Omega(z) + z\,\overline{\Omega'(\bar z)} + \overline{\omega(\bar z)} = \phi(z) \tag{3.90}$$

where $\phi(z)$ is a holomorphic function in the whole plane cut along $L = \bigcup_{r=1}^{n} L_r$. Similarly, as the normal stress $\tau_{yy} - i\tau_{xy}$ is continuous across $y = 0$ outside of the cracks,

$$\Omega(z) - z\,\overline{\Omega'(\bar z)} - \overline{\omega(\bar z)} = \theta(z) \tag{3.91}$$

where $\theta(z)$ is holomorphic in the whole plane cut along L. Solving equations (3.90) and (3.91) we find

$$(1 + \kappa)\, \Omega(z) = \phi(z) + \theta(z) \tag{3.92}$$

and hence $\Omega(z)$ is holomorphic in the whole plane cut along L. Also $\omega(z)$ may be found from (3.91) in terms of $\phi(z)$, $\theta(z)$. We could now substitute the representations (3.92) and (3.91) (or (3.90)), into (3.89) and find expressions which completely satisfy the conditions of continuity outside of the cracks for the displacements and stresses in terms of $\phi(z)$ and $\theta(z)$. However in view of (3.92) it is just as convenient to express the basic equations in terms of $\Omega(z)$ and $\theta(z)$ giving

$$\begin{aligned} 2\mu D &= \kappa\, \Omega(z) - \Omega(\bar{z}) + (\bar{z} - z)\, \overline{\Omega'(z)} + \theta(\bar{z}) \\ \tau_{yy} - \mathrm{i}\tau_{xy} &= \Omega'(z) + \Omega'(\bar{z}) + (z - \bar{z})\, \overline{\Omega''(z)} - \theta'(\bar{z}) \end{aligned} \tag{3.93}$$

where $\Omega(z)$, $\theta(z)$ are holomorphic in the whole plane cut along L. Similar equations could also be derived following the approach of Muskhelishvili,[43] Section 120 or Milne-Thomson,[41] Section 4.12 in which $\omega(z)$ is written in the form of (3.91), but it is felt that the above approach indicates why such a form is chosen.

The boundary conditions over the cracks now become

$$\left.\begin{aligned} \Omega'^{+}(x) + \Omega'^{-}(x) - \theta'^{-}(x) &= -(f_1(x) + \mathrm{i}g_1(x)) \\ \Omega'^{-}(x) + \Omega'^{+}(x) - \theta'^{+}(x) &= -(f_2(x) + \mathrm{i}g_2(x)) \end{aligned}\right\}(x \in L).$$

Hence on subtracting and adding we find

$$\begin{aligned} \theta'^{+}(x) - \theta'^{-}(x) &= f_2(x) - f_1(x) + \mathrm{i}\{g_2(x) - g_1(x)\} = \alpha(x) \\ [2\,\Omega'(x) - \theta'(x)]^{+} &+ [2\,\Omega'(x) - \theta'(x)]^{-} \\ &= -[f_1(x) + f_2(x) + \mathrm{i}\{g_1(x) + g_2(x)\}] = \beta(x) \end{aligned} \tag{3.94}$$

for $x \in L$.

These are two simple Hilbert problems for the determination of $\theta(z)$ and $2\,\Omega(z) - \theta(z)$.

Further as we have assumed zero stresses at infinity we may conclude $\Omega'(z)$ and $\omega'(z)$ are of $O(z^{-1})$ as $|z| \to \infty$, and hence $\theta'(z) = O(z^{-1})$ as $|z| \to \infty$. Clearly, following the results of Section 1.6, the functions $\theta'(z)$, $2\,\Omega'(z) - \theta'(z)$ may be expressed as Cauchy integrals of the type (1.34). In fact (3.94) have the general solutions

$$\begin{aligned} \theta'(z) &= \frac{1}{2\pi\mathrm{i}} \int_L \frac{\alpha(t)\,\mathrm{d}t}{t - z} + Q(z) \\ 2\,\Omega'(z) - \theta'(z) &= \frac{X(z)}{2\pi\mathrm{i}} \int_L \frac{\beta(t)\,\mathrm{d}t}{X^{+}(t)(t - z)} + P(z)X(z) \end{aligned} \tag{3.95}$$

where $X(z)$ has been defined in (1.35) with $\gamma = \frac{1}{2}$ and is such that $z^n X(z) \to 1$ as $|z| \to \infty$. From the known behaviour of $\theta'(z)$ and $\Omega'(z)$ at infinity we see that the polynomial $Q(z) \equiv 0$ and the polynomial $P(z)$ is of degree $(n - 1)$. Thus the complete solution is determined when the n coefficients of $P(z)$ have been found. As the infinite plane contains n holes (the cracks) these constants must be chosen so that the displacement field is single valued, i.e. non-dislocational (see Section 2.9) and hence from (3.93) $\kappa\,\Omega(z) - \Omega(\bar{z}) + \theta(\bar{z})$ must be a single-valued function. Thus for every contour C in the body

$$\int_C \kappa\, \Omega'(z)\, \mathrm{d}z + \int_C \theta'(\bar{z}) - \Omega'(\bar{z})\, \mathrm{d}\bar{z} = 0.$$

In particular if we take a contour C_k surrounding the kth crack and let this contour shrink to a lacet around the crack the potentials must satisfy the condition

$$\int_{C_k} \kappa\, \Omega'(z)\, \mathrm{d}z + \int_{C_k} \theta'(\bar{z}) - \Omega'(\bar{z})\, \mathrm{d}\bar{z}$$

$$= \int_{L_k} (1+\kappa)\{\Omega'^{+}(x) - \Omega'^{-}(x)\} - \{\theta'^{+}(x) - \theta'^{-}(x)\}\, \mathrm{d}x = 0$$

so that from (3.94)

$$(1+\kappa)\int_{L_k} \Omega'^{+}(x) - \Omega'^{-}(x)\, \mathrm{d}x = -\mathrm{i}(X_k + \mathrm{i}\,Y_k) \tag{3.96}$$

where $X_k + \mathrm{i}\,Y_k$ is the resultant force on the kth crack (compare with (2.57)). Applying this condition for $k = 1, 2, \ldots, n$ yields n linear equations for the determination of the n coefficients and thus the solution may be completed.

Problems involving known displacements or other boundary conditions on the cracks may be solved in a similar manner using the general representation (3.93). We now examine a particular problem.

1. *Single crack*

We consider a single crack lying on the interval $(-a, a)$ and opened by equal and opposite pressures $p(x)$ and shears $s(x)$. The equations (3.94) become

$$\left.\begin{aligned} \theta'^{+}(x) - \theta'^{-}(x) &= 0 \\ \{2\,\Omega'(x) - \theta'(x)\}^{+} + \{2\,\Omega'(x) - \theta'(x)\}^{-} &= -2\{p(x) + \mathrm{i}s(x)\} \end{aligned}\right\}(|x| \leqslant a). \tag{3.97}$$

From (3.95) we find $\theta'(z) \equiv 0$ and $\Omega'(z)$ has the form

$$\Omega'(z) = -\frac{1}{2\pi\mathrm{i}X(z)}\int_{-a}^{a} \frac{X^{+}(t)\{p(t) + \mathrm{i}s(t)\}\,\mathrm{d}t}{t - z} + \frac{P(z)}{X(z)} \tag{3.98}$$

where

$$X(z) = (z^2 - a^2)^{1/2} \tag{3.99}$$

and we select the branch such that $z^{-1}X(z) \to 1$ as $|z| \to \infty$. Also we have seen that in general $P(z)$ is a polynomial of degree $(n-1)$ and so in this case is a constant. However the resultant force over the crack is zero and so $\Omega'(z) = O(z^{-2})$ for large $|z|$. Hence we can conclude $P(z) \equiv 0$. This could also be shown from (3.96). The problem has now reduced to performing the line integral in (3.98) and as in the previous sections this may be done by replacing the line integral by a contour integral on a lacet (see Section 1.8).

In particular when a uniform pressure p and shear s act on the crack we find

$$\Omega'(z) = -\tfrac{1}{2}(p + \mathrm{i}s)\left[1 - \frac{z}{(z^2 - a^2)^{1/2}}\right]$$

so that

$$\Omega(z) = -\tfrac{1}{2}(p + \mathrm{i}s)\{z - (z^2 - a^2)^{1/2}\}. \tag{3.100}$$

Now by using equations (3.93) the stress and displacement fields may be evaluated throughout the body. Again care must be taken in calculating the appropriate values of the function $X(z) = (z^2 - a^2)^{1/2}$. Putting $z - a = R_1 e^{\mathrm{i}\theta_1}$, $z + a = R_2 e^{\mathrm{i}\theta_2}$ it is simple to show on $y = 0$

$$X^+(x) = -X^-(x) = \mathrm{i}(a^2 - x^2)^{1/2} \quad \text{for } |x| \leqslant a$$

and $X(x) = \pm(x^2 - a^2)^{1/2}$ for $x > a$ and $x < -a$ respectively. Hence the stress distribution on $y = 0$ outside of the crack is

$$\tau_{yy} - \mathrm{i}\tau_{xy} = (p + \mathrm{i}s)\left\{\frac{|x|}{(x^2 - a^2)^{1/2}} - 1\right\} \quad (|x| \geqslant a)$$

and the displacements of the upper and lower surfaces of the crack are

$$4\mu D = -(p + \mathrm{i}s)\{(\kappa - 1)x \mp \mathrm{i}(\kappa + 1)(a^2 - x^2)^{1/2}\}$$
$$\text{on } y = 0\pm, \ |x| \leqslant a.$$

It may be seen that the uniform stress distribution inflates the crack into an elliptical shape and gives rise to infinite stresses at the tips of the crack.

Many problems involving bodies with cracks in different physical situations have been considered using various methods of solution. A survey of some of these results with a large bibliography has been given by Paris and Sih.[48] In connection with the present section, solutions for an infinite number of cracks equally spaced along a line have been given by Westergaard,[72] Koiter[31] and England and Green[12] and for two cracks by Willmore.[76]

3.11 Alternative methods of solution

Several alternative methods of solution to the class of half-plane problems discussed in this chapter exist. Of these perhaps the most common is the Fourier transform method which is the analogue of the transform methods used for the corresponding problems in three dimensions. The technique is based on the Fourier integral theorem in its complex formulation. If $\Omega'(z)$ is holomorphic for $z \in S^+$ (i.e. $y > 0$) and $O(z^{-1})$ as $|z| \to \infty$ then the integral

$$A(\xi) = \int_{-\infty}^{\infty} \Omega'(z)\, e^{\mathrm{i}\xi z}\, dz \tag{3.101}$$

evaluated on $z = 0+$, is convergent for all real ξ and is such that

$$A(\xi) = 0 \quad (\xi > 0) \tag{3.102}$$

and has the inverse

$$\Omega'(z) = \frac{1}{2\pi}\int_{-\infty}^{\infty} A(\xi)\,\mathrm{e}^{-\mathrm{i}\xi z}\,\mathrm{d}\xi. \tag{3.103}$$

Similar representations hold for $\omega'(z)$, $z \in S^+$ namely

$$\omega'(z) = \frac{1}{2\pi}\int_{-\infty}^{\infty} B(\xi)\,\mathrm{e}^{-\mathrm{i}\xi z}\,\mathrm{d}\xi, \qquad B(\xi) = \int_{-\infty}^{\infty} \omega'(z)\,\mathrm{e}^{\mathrm{i}\xi z}\,\mathrm{d}z \tag{3.104}$$

where

$$B(\xi) = 0 \quad (\xi > 0).$$

The integral representations (3.103) and (3.104) allow the problems of Sections 3.4, 3.6, 3.8, 3.10 to be reduced to, at most, dual integral equations of which the solutions are known. As a simple example we examine the stress boundary value problem of Section 3.4. If on $y = 0+$

$$\tau_{yy} - \mathrm{i}\tau_{xy} = -\{p(x) + \mathrm{i}s(x)\}$$

then

$$\Omega'^{+}(x) + \overline{\Omega'^{+}(x)} + x\,\overline{\Omega''^{+}(x)} + \overline{\omega'^{+}(x)} = -(p(x) + \mathrm{i}s(x)). \tag{3.105}$$

On multiplying (3.105) by $\mathrm{e}^{\mathrm{i}\xi x}$ and integrating along the real axis, the boundary conditions become

$$A(\xi) + \xi\,\overline{A'(-\xi)} + \overline{B(-\xi)} = H(\xi) = -\int_{-\infty}^{\infty} (p(x) + \mathrm{i}s(x))\,\mathrm{e}^{\mathrm{i}\xi x}\,\mathrm{d}x.$$

Hence for $\xi < 0$,

$$A(\xi) = H(\xi)$$

and for $\xi > 0$,

$$\overline{B(-\xi)} = H(\xi) - \xi\,\overline{A'(-\xi)}$$

which implies

$$B(\xi) = \overline{H(-\xi)} + \xi\,H'(\xi) \quad \text{for } \xi < 0.$$

Thus

$$\left.\begin{aligned} \Omega'(z) &= \frac{1}{2\pi}\int_{-\infty}^{0} H(\xi)\,\mathrm{e}^{-\mathrm{i}\xi z}\,\mathrm{d}\xi \\ \omega'(z) &= \frac{1}{2\pi}\int_{-\infty}^{0} [\overline{H(-\xi)} + \xi\,H'(\xi)]\,\mathrm{e}^{-\mathrm{i}\xi z}\,\mathrm{d}\xi. \end{aligned}\right\} \tag{3.106}$$

Particular solutions may now be evaluated.

An account of the use of Fourier transforms in elasticity has been given by Sneddon[54] and more recently Sneddon has reviewed the general methods and solutions of the resulting integral equations.[55] The transforms are most easily evaluated by use of tables compiled by the Bateman Manuscript Project.[14] See also Sneddon and Lowengrub[58] and, for the appropriate mathematical theorems, Titchmarsh.[67] Very many papers have been published in this field and references may be found in the books cited above and in the review by Teodorescu.[63]

A second method of solution follows from noting that as $\Omega'(z)$ is holomorphic in S^+ and bounded at infinity it may be expressed as a Cauchy principal value integral along the boundary $y = 0$ of S^+ of the form†

$$\Omega'(z) = \frac{1}{2\pi i}\int_{-\infty}^{\infty} \frac{\Omega'(t)\,dt}{t - z} + \tfrac{1}{2}\,\Omega'(\infty) \quad (z \in S^+).$$

In addition it may be shown that

$$\frac{1}{2\pi i}\int_{-\infty}^{\infty} \frac{\Omega'(t)\,dt}{t - z} = -\tfrac{1}{2}\,\Omega'(\infty) \quad (z \in S^-)$$

with similar results holding for $\omega'(z)$. If these representations are substituted into the boundary conditions on C (for example (2.68) or (2.69) in differentiated form or some combination of these) the basic boundary value problems reduce to the solution of certain singular integral equations. For the stress and displacement boundary value problems these integral equations have reasonably simple solutions. This method of solution by representing a function holomorphic in a simply connected region S as a Cauchy integral of its boundary values on the boundary C of S and then converting the elastic boundary value problem into a singular integral equation is quite general and avoids the use of particular analytic continuations. It will be referred to as the direct method of solution and is illustrated in detail for the simplest case of circular regions in Section 4.3 and for more general simply connected regions in Section 5.3. Further illustrations have been given by Muskhelishvili,[43] Section 93, Mikhlin,[39] Galin[17] and Sokolnikoff.[59]

Other methods of solution have included particular integral representations for special problems, see for example England and Green,[12] Williams.[74].

Examples on Chapter 3

1. Determine the stress and displacement fields in an infinite medium due to

† See Muskhelishvili,[43] Section 72.

equal and opposite point forces acting at different points along their common line of action. Show that the displacement at infinity is bounded.

2. Consider the effect of a point force $X + \mathrm{i}Y$ acting at an internal point z_0 of the half-plane S^+ when the boundary $y = 0$ is unstressed. Determine the normal displacement on $y = 0$.

3. If the boundary $y = 0$ of the half-plane S^+ is held so that $\tau_{xy} = 0$ and $v = 0$ find the stress field in S^+ due to the point force $X + \mathrm{i}Y$ acting at a point z_0 in S^+. Determine the normal stress on $y = 0$. (Hint: consider an infinite plate containing the given point force and its image $X - \mathrm{i}Y$ at $\bar{z}_0$.)

4. By using the method of Section 3.5 and the displacement continuation described in Section 3.3 show that the complex potentials due to a system of point forces and moments in the half-plane S^+ are

$$\Omega(z) = \Omega_0(z) + \frac{z}{\kappa}\overline{\Omega_0'(\bar{z})} + \frac{1}{\kappa}\overline{\omega_0(\bar{z})},$$

$$\omega(z) = \omega_0(z) - \frac{z}{\kappa}\overline{\omega_0'(\bar{z})} + \kappa\,\overline{\Omega_0(\bar{z})} - \frac{z}{\kappa}\overline{\Omega_0'(\bar{z})} - \frac{z^2}{\kappa}\overline{\Omega_0''(\bar{z})}$$

when the complex displacement $D \equiv 0$ on $y = 0$.

5. In the case of a flat-ended rough punch normally indenting the half-plane S^+ show that the displacement gradient $\partial D/\partial x$ on $y = 0$ oscillates infinitely often as $|x| \to a+$. Confirm that the length of this region of oscillation is approximately the same as the length of the region of stress oscillation.

6. Using the theory of Section 3.7 consider the situation of two flat-ended rough punches normally indenting the half-plane S^+. If the punches make contact with S^+ over the intervals $-b \leqslant x \leqslant -a$, $a \leqslant x \leqslant b$ and the resultant forces on the punches are P_1, P_2 show that

$$\Omega'(z) = (D_0 + D_1 z)X(z)$$

where $X(z)$ is the appropriate Plemelj function

$$X(z) = (z + b)^{-\gamma}(z + a)^{\gamma-1}(z - a)^{-\gamma}(z - b)^{\gamma-1}, \qquad \gamma = \tfrac{1}{2} + \frac{\mathrm{i}}{2\pi}\log\kappa$$

and

$$D_1 = -(P_1 + P_2)/2\pi$$

$$D_0\alpha = (P_1 + P_2)\frac{\beta}{2\pi} - \frac{P_2}{\mathrm{i}(1 + \kappa)}$$

where

$$\alpha = \int_a^b X^+(x)\,dx, \qquad \beta = \int_a^b xX^+(x)\,dx.$$

7. Using the basic formulae (2.44) show that for a medium occupying the half-plane S^+ the boundary condition $\tau_{xy} = 0$ on $y = 0+$ implies that the functions $z\,\Omega''(z) + \omega'(z)$ and $z\,\overline{\Omega''(\bar{z})} + \overline{\omega'(\bar{z})}$ continue each other analytically across $y = 0$. Hence conclude $\omega'(z) = -z\,\Omega''(z)$ for $z \in S^+$. Show that if in addition the normal stress $\tau_{yy} = f(x)$ is specified on $y = 0$ then the holomorphic function $\theta'(z)$ defined by

$$\theta'(z) = \Omega'(z) \quad (z \in S^+), \qquad \theta'(z) = -\overline{\Omega'(\bar{z})} \quad (z \in S^-)$$

is such that

$$\theta'^+(x) - \theta'^-(x) = f(x) \quad \text{on } y = 0.$$

Hence express $\theta'(z)$ as a Cauchy integral and confirm that the necessary condition $\Omega'(z) = \theta'(z) = -\overline{\theta'(\bar{z})}$ for $z \in S^+$ is satisfied. Investigate the solution when

$$f(x) = -p_0 \quad (|x| \leqslant a), \qquad f(x) = 0 \quad (|x| > a).$$

This method may also be applied to problem 3.

8. Solve the problem of a frictionless punch acting on the half-space S^+ by using the boundary condition $\tau_{xy} = 0$ to introduce the function $\theta'(z)$ as in the previous example. Confirm that $\theta'(z)$ is holomorphic in the whole plane cut along the contact region and satisfies a Hilbert problem. Determine the solution in the case of a flat-ended punch acting normal to the half-space.

9. Determine the stress distribution under a smooth wedge-shaped punch having an end section $y = -\varepsilon x \quad (0 \leqslant x \leqslant l)$, which is in partial contact with the half-plane S^+. What is the minimum force which should be applied to the punch to ensure complete contact with S^+?

10. Extend the theory of Section 3.10 to include the case of known bounded stresses at infinity. Hence determine the stress field around an unstressed line crack in an infinite medium under a uniform tension T at infinity acting at an inclination ϕ to the line of the crack. Confirm that

$$\Omega'(z) = \frac{T}{4}e^{2i\phi} + \frac{T}{4}(1 - e^{2i\phi})\frac{z}{(z^2 - a^2)^{1/2}}, \qquad \theta'(z) = \frac{T}{2}e^{2i\phi}.$$

11. A composite material occupying the region R is formed of two dissimilar elastic materials. If the two materials are bonded together (stress and displacement continuous) along the part of $y = 0$ in R except over a region

L of $y = 0$, show that the complex potentials $\Omega_1(z)$, $\omega_1(z)$ and $\Omega_2(z)$, $\omega_2(z)$ in the two media satisfy the equations

$$(\mu_1 + \mu_2\kappa_1)\,\Omega_1(z) = \mu_1\theta(z) + \phi(z)$$

$$(\mu_2 + \mu_1\kappa_2)\,\Omega_2(z) = \mu_2\theta(z) + \phi(z)$$

$$\omega_1(z) = \overline{\Omega_2(\bar{z})} - z\,\Omega_1'(z) - \overline{\theta(\bar{z})}$$

$$\omega_2(z) = \overline{\Omega_1(\bar{z})} - z\,\Omega_2'(z) - \overline{\theta(\bar{z})}$$

where $\theta(z)$, $\phi(z)$ are holomorphic functions in the region R cut along L and $\mu_1, \kappa_1; \mu_2, \kappa_2$ are the elastic constants in the two media.

12. Use the above results to consider the deformation of the infinite body composed of the medium with elastic constants μ_1, κ_1 occupying S^+ $(y > 0)$ and bonded to the medium with elastic constants μ_2, κ_2 occupying S^- $(y < 0)$ under the action of the stresses $\tau_{xx} = X_1$, $\tau_{yy} = Y_1$ in S^+ and $\tau_{xx} = X_2$, $\tau_{yy} = Y_2$ in S^-. Examine the displacement on $y = 0$.

13. Suppose two dissimilar elastic media occupy the half-planes S^+ and S^- and are bonded together (displacement and stress continuous) along $y = 0$ except over the region $|x| \leqslant a$ where there is a line crack. If this crack is opened by equal and opposite normal pressures $p(x)$ on each side of the crack, by using the results of example 11 show that $\theta(z)$ must be such that

$$\theta'^+(x) - \theta'^-(x) = 0 \quad \text{for all } x \text{ on } y = 0.$$

Hence conclude that $\theta(z) \equiv 0$.

Show that $\phi(z)$ satisfies the Hilbert problem

$$\phi'^+(x) + \alpha\phi'^-(x) = -(\mu_1 + \mu_2\kappa_1)p(x)$$

where $\alpha = (\mu_1 + \mu_2\kappa_1)/(\mu_2 + \mu_1\kappa_2)$.

Confirm that when $p(x)$ is constant

$$\theta(z) = 0, \qquad \phi(z) = K\left[z - \left(\frac{z + a}{z - a}\right)^{i\beta}(z^2 - a^2)^{1/2}\right]$$

where K is a constant and $2\pi\beta = \log \alpha$. Hence show that the distance apart of the surfaces of the crack is proportional to

$$(a^2 - x^2)^{1/2}\cos(\beta \log|(x + a)/(x - a)|)$$

which indicates that the crack surfaces overlap near the ends of the crack. Show also that the stresses have an essential singularity at the ends of the crack.

4

REGIONS WITH CIRCULAR BOUNDARIES

Introduction

In this chapter we shall examine the various methods of solution of some elastic boundary value problems for a circular disc, an infinite medium containing a circular hole, or an annular region (the region between two concentric circles). In general there are three basic methods of solution: the first, using Laurent's theorem to express each complex potential as a power series in the above regions, reduces the stress and displacement boundary value problems to sets of simultaneous linear equations in the (unknown) coefficients of the two power series. We shall refer to this as the *series method* of solution. The second method of solution is based on the properties of Cauchy integrals on the circle and yields a very convenient solution to the stress and displacement problems. This will be referred to as the *direct method* of solution. The above methods are not immediately applicable to more general boundary value problems such as the mixed problem, however the method based on continuation of one of the complex potentials across the circular boundary, which was illustrated in the case of a straight-line boundary in the last chapter, is much more powerful and allows several types of boundary value problems to be solved with comparative ease.

For convenience we repeat at this point the relations with the complex potentials in polar coordinates from (2.74)

$$\begin{aligned} 2\mu(u_r + \mathrm{i}u_\theta) &= 2\mu D\mathrm{e}^{-\mathrm{i}\theta} = \mathrm{e}^{-\mathrm{i}\theta}\{\kappa\,\Omega(z) - z\,\overline{\Omega'(z)} - \overline{\omega(z)}\} \\ \tau_{rr} + \mathrm{i}\tau_{r\theta} &= \Omega'(z) + \overline{\Omega'(z)} - \left(\bar{z}\,\overline{\Omega''(z)} + \frac{\bar{z}}{z}\,\overline{\omega'(z)}\right) \\ \tau_{rr} + \tau_{\theta\theta} &= 2\{\Omega'(z) + \overline{\Omega'(z)}\}. \end{aligned} \tag{4.1}$$

We shall always take the origin at the centre of the circular boundaries.

4.1 Stress and displacement boundary value problems

In this section we show that both the stress and displacement boundary value problems for a region with a general boundary contour C reduce to finding single-valued potentials which satisfy the boundary condition

$$\gamma\, \Omega(t) + t\,\overline{\Omega'(t)} + \overline{\omega(t)} = F(t)$$

for all points t of C.

As discussed in Section 2.11, the displacement boundary value problem for a body with general boundary contour C amounts to specifying $u + \mathrm{i}v = D(t)$ at all points of C. It will be seen from (4.1) that this is equivalent to the boundary condition.

$$\kappa\, \Omega(t) - t\,\overline{\Omega'(t)} - \overline{\omega(t)} = 2\mu\, D(t) \quad \text{on } C \tag{4.2}$$

where for a non-dislocational displacement field $D(t)$ is a known single-valued function on C.

In the case of a simply connected region it was shown in Section 2.9 that the complex potentials are single valued, this however is not true for more general regions. As we are concerned in this chapter with bodies having at most one internal hole, if we take the origin in the interior of the hole, the complex potentials have the form, from (2.57),

$$\begin{aligned} \Omega(z) &= -\frac{X + \mathrm{i}Y}{2\pi(1+\kappa)} \log z + \Omega_0(z), \\ \omega(z) &= \frac{\kappa(X - \mathrm{i}Y)}{2\pi(1+\kappa)} \log z + \omega_0(z) \end{aligned} \tag{4.3}$$

where $X + \mathrm{i}Y$ is the resultant force over the hole and $\Omega_0(z)$, $\omega_0(z)$ are single-valued holomorphic functions in the body. On substituting in (4.2) the boundary condition for $\Omega_0(z)$, $\omega_0(z)$ becomes

$$\begin{aligned} \kappa\, \Omega_0(t) - t\,\overline{\Omega_0'(t)} - \overline{\omega_0(t)} = 2\mu\, D(t) &+ \frac{\kappa(X + \mathrm{i}Y)}{2\pi(1+\kappa)} \log(t\bar{t}) \\ &- \frac{(X - \mathrm{i}Y)\,t}{2\pi(1+\kappa)\,\bar{t}}. \end{aligned} \tag{4.4}$$

It will be seen that all functions in (4.4) are single valued and the right-hand side is a known function, except that the value of $X + \mathrm{i}Y$ must be determined after the solution is obtained. Thus even for a doubly connected region the boundary conditions are of the form (4.2) and involve only single-valued functions.

Similarly in the case of the stress boundary value problem the resultant force $R(t)$ over an arc AT of the boundary (taken in the direction so that the body lies to the left of AT, see Figure 2.3) is, from (2.47),

$$R(t) = -\mathrm{i}[\Omega(z) + z\,\overline{\Omega'(z)} + \overline{\omega(z)}]_A^T$$

where A is a fixed point on the boundary. Hence the boundary conditions for the stress problem become

$$\Omega(t) + t\,\overline{\Omega'(t)} + \overline{\omega(t)} = \mathrm{i}\,R(t) + \text{constant}. \tag{4.5}$$

However, in the case of the stress boundary value problem the displacement field is undetermined to within a rigid-body displacement, and using the results of Section 2.8 this is equivalent to the left-hand side of (4.5) being undetermined to within an arbitrary constant. Thus for the stress boundary value problem it is sufficient to determine complex potentials satisfying

$$\Omega(t) + t\,\overline{\Omega'(t)} + \overline{\omega(t)} = \mathrm{i}\,R(t) \quad \text{on } C \tag{4.6}$$

where $R(t)$ is the (known) resultant force over an arc of the boundary measured from some fixed point. In terms of the boundary values $R(t)$ has the form $-\mathrm{i}\int_{t_0}^{t}(\tau_{rr} + \mathrm{i}\tau_{r\theta})\,\mathrm{d}t$ where the direction of integration keeps S^+ to the left.

Again this boundary condition must be examined in detail in the cases of simply and doubly connected regions. For a simply connected region it has been shown in Section 2.9 that the complex potentials are single valued, and hence from (4.6) $R(t)$ must be single valued in a circuit of C. This is equivalent to the necessary physical condition that the resultant force on the body be zero. Similarly, as the resultant moment on the body must also be zero, $R(t)$ is such that†

$$\mathrm{Im}\int_C R(t)\,\mathrm{d}\bar{t} = 0. \tag{4.7}$$

For a doubly connected region, since the complex potentials have the form (4.3) where $X + \mathrm{i}Y$ is the known resultant force over the hole the boundary condition becomes

$$\Omega_0(t) + t\,\overline{\Omega_0'(t)} + \overline{\omega_0(t)} = \mathrm{i}\,R(t) + \frac{X + \mathrm{i}Y}{2\pi(1+\kappa)}(\log t - \kappa \log \bar{t}) + \frac{(X - \mathrm{i}Y)}{2\pi(1+\kappa)}\frac{t}{\bar{t}} \quad \text{on } C$$

† Using the notation of Section 2.7, the resultant moment is

$$-\int_C (Py - Qx)\,\mathrm{d}s = \mathrm{Im}\int_C (P + iQ)\bar{t}\,\mathrm{d}s$$

and on integrating by parts

$$= \mathrm{Im}\left\{[R(t)\bar{t}]_C - \int_C R(t)\frac{\mathrm{d}\bar{t}}{\mathrm{d}s}\,\mathrm{d}s\right\} = 0.$$

Since the first term is zero, (4.7) results.

and on putting $t = re^{i\theta}$ on C

$$\Omega_0(t) + t\,\overline{\Omega_0'(t)} + \overline{\omega_0(t)} = \mathrm{i}\,R(re^{i\theta}) + \frac{(X+\mathrm{i}Y)}{2\pi}\mathrm{i}\theta + \frac{(X+\mathrm{i}Y)(1-\kappa)}{2\pi(1+\kappa)}\log r + \frac{X-\mathrm{i}Y}{2\pi(1+\kappa)}e^{2i\theta}$$

$$= \mathrm{i}\,R_0(t) \tag{4.8}$$

where $R_0(t)$ is a known function on C. Thus if C is the internal boundary of the doubly connected region, as C is described once in a clockwise manner, $R(t)$ changes by $X + \mathrm{i}Y$ and θ by -2π and so $R_0(t)$ is a single-valued function on C. Similarly, if the doubly connected region is bounded and C is the external contour, in a single anti-clockwise circuit of C, $R(t)$ increases by $-(X + \mathrm{i}Y)$ and θ by 2π and again $R_0(t)$ is single valued. Thus, on omitting the subscripts in (4.8), for both simply and doubly connected regions the stress boundary value problem reduces to the boundary conditions

$$\Omega(t) + t\,\overline{\Omega'(t)} + \overline{\omega(t)} = \mathrm{i}\,R(t) \quad \text{on } C \tag{4.9}$$

where all functions are single valued in their regions of definition.

On comparing equations (4.2) and (4.9) it is clear that both the stress and displacement boundary value problems may be solved by examining the equation

$$\gamma\,\Omega(t) + t\,\overline{\Omega'(t)} + \overline{\omega(t)} = F(t) \quad \text{on } C \tag{4.10}$$

where $\gamma = 1$ for the stress problem, $\gamma = -\kappa$ for the displacement problem, $\Omega(z)$, $\omega(z)$ are single-valued holomorphic functions and the single-valued function $F(t)$ is interpreted according to equations (4.2), (4.4), (4.6) or (4.8).

The next four sections illustrate the three basic methods of solution of the stress and displacement boundary value problems for regions with circular boundaries using the boundary conditions in the form (4.10). It should be noted that a similar approach could have been adopted with the half-plane problems discussed in Chapter 3.

4.2 Series method of solution

As a representative example of this method of solution we shall examine the case of an annulus $a \leqslant |z| \leqslant b$ with the boundary conditions

$$\begin{aligned} \gamma_1\,\Omega(t) + t\,\overline{\Omega'(t)} + \overline{\omega(t)} &= F_1(t) \quad \text{on } t = ae^{i\theta} \\ \gamma_2\,\Omega(t) + t\,\overline{\Omega'(t)} + \overline{\omega(t)} &= F_2(t) \quad \text{on } t = be^{i\theta}. \end{aligned} \tag{4.11}$$

As has been pointed out in the last section the stress and displacement boundary value problems may be reduced to determining functions $\Omega(z)$, $\omega(z)$ which are holomorphic and single valued in the annulus and satisfy conditions (4.11). By Laurent's theorem (Section 1.3) the complex potentials may be represented by the power series

$$\Omega(z) = \sum_{-\infty}^{\infty} \alpha_n z^n, \qquad \omega(z) = \sum_{-\infty}^{\infty} \beta_n z^n \tag{4.12}$$

for $a \leqslant |z| \leqslant b$. On substituting in (4.11) the boundary conditions become

$$\gamma_1 \sum_{-\infty}^{\infty} \alpha_n t^n + t \sum_{-\infty}^{\infty} n\bar{\alpha}_n \bar{t}^{n-1} + \sum_{-\infty}^{\infty} \bar{\beta}_n \bar{t}^n = F_1(t) \quad \text{on } t = a\mathrm{e}^{\mathrm{i}\theta}$$

$$\gamma_2 \sum_{-\infty}^{\infty} \alpha_n t^n + t \sum_{-\infty}^{\infty} n\bar{\alpha}_n \bar{t}^{n-1} + \sum_{-\infty}^{\infty} \bar{\beta}_n \bar{t}^n = F_2(t) \quad \text{on } t = b\mathrm{e}^{\mathrm{i}\theta}.$$

Following the usual techniques of Fourier series, on multiplying through each equation by $\mathrm{e}^{-\mathrm{i}p\theta}$ and integrating with respect to θ from $-\pi$ to π, the problem reduces to the solution of the following sets of simultaneous linear equations

$$\begin{aligned} \gamma_1 a^{2p}\alpha_p + (2-p)a^2\bar{\alpha}_{2-p} + \bar{\beta}_{-p} &= \frac{a^p}{2\pi}\int_{-\pi}^{\pi} F_1(a\mathrm{e}^{\mathrm{i}\theta})\mathrm{e}^{-\mathrm{i}p\theta}\,\mathrm{d}\theta \\ \gamma_2 b^{2p}\alpha_p + (2-p)b^2\bar{\alpha}_{2-p} + \bar{\beta}_{-p} &= \frac{b^p}{2\pi}\int_{-\pi}^{\pi} F_2(b\mathrm{e}^{\mathrm{i}\theta})\mathrm{e}^{-\mathrm{i}p\theta}\,\mathrm{d}\theta \end{aligned} \tag{4.13}$$

for integral values of p in the range $(-\infty, \infty)$. On subtracting these equations the constants α_n satisfy

$$(\gamma_1 a^{2p} - \gamma_2 b^{2p})\alpha_p + (2-p)(a^2 - b^2)\bar{\alpha}_{2-p} = c_p \tag{4.14}$$

and then the β_n are given by

$$(a^{-2} - b^{-2})\beta_p = (\gamma_2 b^{-2(p+1)} - \gamma_1 a^{-2(p+1)})\bar{\alpha}_{-p} + d_p \tag{4.15}$$

where

$$\begin{aligned} c_p &= \frac{1}{2\pi}\int_{-\pi}^{\pi} \{a^p F_1(a\mathrm{e}^{\mathrm{i}\theta}) - b^p F_2(b\mathrm{e}^{\mathrm{i}\theta})\}\mathrm{e}^{-\mathrm{i}p\theta}\,\mathrm{d}\theta \\ d_p &= \frac{1}{2\pi}\int_{-\pi}^{\pi} \{a^{-p-2}\,\overline{F_1(a\mathrm{e}^{\mathrm{i}\theta})} - b^{-p-2}\,\overline{F_2(b\mathrm{e}^{\mathrm{i}\theta})}\}\mathrm{e}^{-\mathrm{i}p\theta}\,\mathrm{d}\theta. \end{aligned} \tag{4.16}$$

Finally on replacing p in (4.14) by $2 - p$ and taking the conjugate of the resulting equation it is possible to evaluate α_p and find

$$\Delta_p\alpha_p = (\gamma_1 a^{2(2-p)} - \gamma_2 b^{2(2-p)})c_p - (2-p)(a^2 - b^2)\bar{c}_{2-p} \tag{4.17}$$

where

$$\Delta_p = (\gamma_1 a^{2p} - \gamma_2 b^{2p})(\gamma_1 a^{2(2-p)} - \gamma_2 b^{2(2-p)}) - p(2-p)(a^2 - b^2)^2. \quad (4.18)$$

Clearly the cases where $\Delta_p = 0$ require special consideration.

In the stress boundary value problem, when $\gamma_1 = \gamma_2 = 1, \Delta_p$ is non-zero except when $p = 0$, $p = 1$, $p = 2$. For $p = 0$, $p = 2$ equations (4.14) become

$$2(a^2 - b^2)\bar{\alpha}_2 = c_0, \qquad (a^4 - b^4)\alpha_2 = c_2 \quad (4.19)$$

and α_0 remains undefined as expected, since it corresponds to a rigid-body translation. Hence for consistency we require

$$(a^2 + b^2)\bar{c}_0 = 2c_2. \quad (4.20)$$

It will be recalled from the definition of $F_1(t)$, $F_2(t)$ that these functions contain arbitrary constants. These constants contribute to c_0 and hence the consistency condition (4.20) may be satisfied. This slight difficulty is examined in the first illustration. In the third case when $p = 1$ equation (4.14) becomes

$$(a^2 - b^2)(\alpha_1 + \bar{\alpha}_1) = c_1 \quad (4.21)$$

and hence c_1 must be real. It is simple to confirm from equations (4.7) and (4.8) that $\text{Im}(c_1)$ is proportional to the resultant moment applied to the annulus and so must be zero. In addition only the real part of α_1 is specified in the solution, which is consistent with the displacement field being determined only to within a rigid-body motion.

For the displacement boundary value problem when $\gamma_1 = \gamma_2 = -\kappa$, Δ_p is non zero except when $p = 0$ and $p = 2$. In these cases equations (4.14) become

$$2(a^2 - b^2)\bar{\alpha}_2 = c_0, \qquad -\kappa(a^4 - b^4)\alpha_2 = c_2$$

and the consistency equation is

$$\kappa(a^2 + b^2)\bar{c}_0 = -2c_2. \quad (4.22)$$

However in this case if $D_1(t)$, $D_2(t)$ are the applied displacements on $|t| = a$, $|t| = b$ respectively, then from (4.4)

$$F_\alpha(t) = -2\mu\, D_\alpha(t) - \frac{\kappa(X + \mathrm{i}Y)}{\pi(1 + \kappa)} \log|t| + \frac{(X - \mathrm{i}Y)}{2\pi(1 + \kappa)} \mathrm{e}^{2\mathrm{i}\theta} \quad \text{for } \alpha = 1, 2$$

and the consistency equation (4.22) becomes

$$(X + \mathrm{i}Y)\left[1 - \kappa^2 \frac{(b^2 + a^2)}{(b^2 - a^2)} \log\left(\frac{b}{a}\right)\right]$$
$$= \frac{2\mu(1+\kappa)}{b^2 - a^2} \int_{-\pi}^{\pi} \{b^2 \overline{D_2(b\mathrm{e}^{\mathrm{i}\theta})} - a^2 \overline{D_1(a\mathrm{e}^{\mathrm{i}\theta})}\} \mathrm{e}^{2\mathrm{i}\theta}\, \mathrm{d}\theta$$
$$+ \mu\kappa(1+\kappa)\frac{(b^2 + a^2)}{(b^2 - a^2)} \int_{-\pi}^{\pi} \{D_2(b\mathrm{e}^{\mathrm{i}\theta}) - D_1(a\mathrm{e}^{\mathrm{i}\theta})\}\, \mathrm{d}\theta$$

defining the resultant force $X + \mathrm{i}Y$ over the hole.

It will also be noticed that α_0, β_0 are only defined by the equation

$$-\kappa\alpha_0 + \bar{\beta}_0 = \frac{1}{2\pi}\int_{-\pi}^{\pi} F_1(a\mathrm{e}^{\mathrm{i}\theta})\, \mathrm{d}\theta - 2a^2\bar{\alpha}_2$$

however it has been shown in Section 2.8 that only the combination $\kappa\alpha_0 - \bar{\beta}_0$ of the constants α_0, β_0 contributes to the displacement field and hence one of them may be defined arbitrarily (say $\beta_0 = 0$).

Particular solutions to various boundary value problems for the annulus may now be obtained directly from the above equations including the case where the displacement is specified on one boundary, the stress on the other.

Consideration has not been given so far to the cases of a circular disc $|z| \leqslant a$ or an infinite medium $|z| \geqslant a$; however, these solutions may be immediately derived from (4.10), (4.12) and the first equation of (4.13) and are discussed in the following illustrations.

1. *Annulus under uniform internal and external pressures*

Suppose $\tau_{rr} = -p_1$, on $|z| = a$, $\tau_{rr} = -p_2$ on $|z| = b$, then the resultant force $X + \mathrm{i}Y$ over the hole is zero and from (4.8) and (4.10) we have

$$\gamma_1 = \gamma_2 = 1$$

$$F_1(t) = -p_1 t + \delta_1, \qquad F_2(t) = -p_2 t + \delta_2$$

where δ_1, δ_2 are arbitrary constants.

From (4.14)

$$c_0 = \delta_1 - \delta_2, \qquad c_1 = b^2 p_2 - a^2 p_1$$

$$d_0 = \frac{\bar{\delta}_1}{a^2} - \frac{\bar{\delta}_2}{b^2}, \qquad d_{-1} = p_2 - p_1$$

the remaining constants being zero. On substituting into (4.20), (4.19), (4.21), (4.14) and (4.15) we find

$$\delta_1 = \delta_2,$$

$$\alpha_1 = \frac{b^2 p_2 - a^2 p_1}{2(b^2 - a^2)}$$

$$\beta_{-1} = \frac{(p_2 - p_1)a^2 b^2}{b^2 - a^2}$$

the remaining constants being zero and α_0, β_0 undetermined. These yield the following potentials

$$\Omega(z) = \frac{b^2 p_2 - a^2 p_1}{2(a^2 - b^2)} z, \qquad \omega(z) = \frac{(p_2 - p_1)a^2 b^2}{(b^2 - a^2)z}$$

from which the stress and displacement fields may be calculated.

2. *Circular disc*

In the case of a circular disc $|z| \leqslant a$ the Laurent series (4.12) are

$$\Omega(z) = \sum_0^\infty \alpha_n z^n, \qquad \omega(z) = \sum_0^\infty \beta_n z^n$$

and the stress or displacement boundary conditions

$$\gamma\, \Omega(t) + t\, \overline{\Omega'(t)} + \overline{\omega(t)} = F(t) \quad \text{on } t = a\mathrm{e}^{\mathrm{i}\theta}$$

become

$$\gamma a^{2p} \alpha_p + (2 - p)a^2 \bar{\alpha}_{2-p} + \bar{\beta}_{-p} = \frac{a^p}{2\pi} \int_{-\pi}^{\pi} F(a\mathrm{e}^{\mathrm{i}\theta})\mathrm{e}^{-\mathrm{i}p\theta}\, \mathrm{d}\theta = F_p \quad (4.23)$$

from the first equation of (4.13). Remembering that $\alpha_p = \beta_p = 0$ for $p < 0$ equation (4.23) reduces to

$$\gamma a^{2p} \alpha_p = F_p \quad (p \geqslant 2)$$

$$a^2(\gamma \alpha_1 + \bar{\alpha}_1) = F_1$$

$$\gamma \alpha_0 + 2a^2 \bar{\alpha}_2 + \bar{\beta}_0 = F_0$$

$$\beta_p = \bar{F}_{-p} - (2 + p)a^2 \alpha_{2+p} \quad (p > 0).$$

As in the case of the annular region the solution to the stress or displacement problem may be simply found, the solution containing only those arbitrary constants which are consistent with the physical situation. Particular examples of this solution are left to the reader.

3. *Circular hole in an infinite medium*

In this case the region $|z| \geqslant a$ is doubly connected and the boundary condition for the stress or displacement boundary value problem is

$$\gamma\, \Omega(t) + t\, \overline{\Omega'(t)} + \overline{\omega(t)} = F(t), \qquad t = ae^{i\theta}$$

where $F(t)$ is defined by (4.8) or (4.4). The potentials $\Omega(z)$, $\omega(z)$ are holomorphic and single valued in $|z| \geqslant a$ and have Laurent expansions

$$\Omega(z) = (A + iB)z + \sum_0^\infty \alpha_n z^{-n}, \qquad \omega(z) = (C + iD)z + \sum_0^\infty \beta_n z^{-n}$$

where $A + iB$, $C + iD$ have been defined in terms of the stress and rotation at infinity by (2.61), (2.62) and may be regarded as known constants. The boundary condition from (4.13) is then

$$\gamma a^{2p}\alpha_{-p} + (2 - p)a^2\bar{\alpha}_{p-2} + \bar{\beta}_p = F_p$$

where

$$\begin{aligned} F_p &= \frac{a^p}{2\pi}\int_{-\pi}^{\pi} F(ae^{i\theta})e^{-ip\theta}\, d\theta \quad (p \neq \pm 1) \\ F_1 &= \frac{a}{2\pi}\int_{-\pi}^{\pi} F(ae^{i\theta})e^{-i\theta}\, d\theta - a^2\{\gamma(A + iB) + A - iB\} \\ F_{-1} &= \frac{1}{2\pi a}\int_{-\pi}^{\pi} F(ae^{i\theta})e^{i\theta}\, d\theta - (C - iD) \end{aligned} \tag{4.24}$$

and hence

$$\begin{aligned} \gamma\alpha_p &= a^{2p}F_{-p} \quad (p \geqslant 1) \\ \gamma\alpha_0 + \bar{\beta}_0 &= F_0 \\ \beta_1 &= \bar{F}_1 \\ \beta_p &= \bar{F}_p + (p - 2)a^2\bar{\alpha}_{p-2} \quad (p \geqslant 2). \end{aligned} \tag{4.25}$$

Particular solutions to the problem may now be obtained to within the appropriate arbitrary constants.

Consider an infinite plate containing an unstressed circular hole deformed by a known uniaxial stress at infinity $\tau_{xx} = T$, $\tau_{xy} = \tau_{yy} = 0$. Then in the above $A = \frac{1}{4}T$, $B = 0$, $C + iD = -\frac{1}{2}T$ and $\gamma = 1$, $F(t) = 0$. Hence from (4.24) and (4.25)

$$\alpha_1 = \tfrac{1}{2}a^2T, \qquad \alpha_0 + \bar{\beta}_0 = 0$$
$$\beta_1 = -\tfrac{1}{2}a^2T, \qquad \beta_3 = \tfrac{1}{2}a^4T.$$

Thus the complete potentials are

$$\Omega(z) = \frac{T}{4}\left(z + \frac{2a^2}{z}\right), \qquad \omega(z) = -\frac{T}{2}\left(z + \frac{a^2}{z} - \frac{a^4}{z^3}\right). \tag{4.26}$$

As a second example consider a loaded circular hole with zero stress and rotation at infinity. If we suppose the stress distribution

$$\tau_{rr} + i\tau_{r\theta} = -\frac{(X + iY)}{2\pi a} e^{-i\theta}$$

acts over the hole it has the resultant force $X + iY$. Then from (4.8) we see that

$$R(t) = -\frac{(X+iY)\theta}{2\pi}, \quad \text{where } t = ae^{i\theta}$$

d hence

$$F(t) = \frac{(X + iY)(1 - \kappa)}{2\pi(1 + \kappa)} \log a + \frac{(X - iY)}{2\pi(1 + \kappa)} e^{2i\theta}.$$

Thus from (4.24) and (4.25)

$$\beta_2 = \frac{a^2(X + iY)}{2\pi(1 + \kappa)}$$

the remaining constants are either zero or can be neglected and the complete potentials are

$$\Omega(z) = -\frac{(X + iY)}{2\pi(1 + \kappa)} \log z, \qquad \omega(z) = \frac{\kappa(X - iY)}{2\pi(1 + \kappa)} \log z + \frac{(X + iY)\, a^2}{2\pi(1 + \kappa)\, z^2}. \tag{4.27}$$

In the limit as $a \to 0$ these potentials become those associated with a point force $X + iY$ at the origin in an infinite medium (see Section 3.2). In fact a point force may be defined as the limit of the above stress distribution over the circular hole as the radius a tends to zero.

It will be appreciated that the series method of solution forms a straightforward and efficient technique for the solution of simple stress and displacement boundary value problems.† For more complicated boundary conditions the series solution is less effective than the method based on continuation which is described in Section 4.4. If however the series solution is used for some simple mixed boundary conditions the problem reduces to the solution of a system of dual series equations. Some equations of this type have been examined by Cooke and Tranter,[9] Williams,[75] and are discussed by Sneddon,[55] Chapter 5.

† Many problems have been solved by this method see, for example, Love,[38] Timoshenko and Goodier,[66] Chapter 4, Sokolnikoff,[59] Chapter 5, and Muskhelishvili,[43] Part III.

4.3 Direct method of solution

In this section we discuss a method which is based on the properties of Cauchy integrals around a circle and provides a very convenient solution to the stress and displacement boundary value problems for a circular disc or an infinite medium containing a circular hole. The method may be extended to solve these problems for an annular region but this extension is omitted.

We examine first the case of the circular disc $|z| < a$ and denote this region by S^+, the exterior region $|z| > a$ by S^-, and the circle $|z| = a$ by Γ and suppose it to be described in an anticlockwise manner. From Section 4.1, for the stress and displacement boundary value problems, the complex potentials $\Omega(z)$, $\omega(z)$ are single valued and holomorphic in S^+ and satisfy the boundary conditions

$$\gamma\,\Omega(t) + t\,\overline{\Omega'(t)} + \overline{\omega(t)} = F(t) \quad \text{on } \Gamma \tag{4.28}$$

where $F(t)$ is single valued on Γ. Now the function $\omega(z)$ is holomorphic in $|z| \leqslant a$ and has the boundary value $\omega(t)$ on Γ. Hence by the theorems of Section 1.5 we can conclude

$$\frac{1}{2\pi i}\int_\Gamma \frac{\overline{\omega(t)}\,dt}{t - z} = \overline{\omega(0)} \quad \text{for } z \in S^+. \tag{4.29}$$

If we choose $\omega(z)$ so that

$$\omega(0) = 0 \tag{4.30}$$

which is possible in both the stress and displacement problems, then from (4.28) and (4.29)

$$\frac{1}{2\pi i}\int_\Gamma \frac{\{F(t) - \gamma\,\Omega(t) - t\,\overline{\Omega'(t)}\}\,dt}{t - z} = 0 \quad \text{for } z \in S^+. \tag{4.31}$$

Now, as $\Omega(z)$ is holomorphic in S^+ from Section 1.5,

$$\frac{1}{2\pi i}\int_\Gamma \frac{\Omega(t)\,dt}{t - z} = \Omega(z) \quad (z \in S^+) \tag{4.32}$$

and we can also evaluate the integral involving $\overline{\Omega'(t)}$.

Since $\Omega(z)$ is holomorphic in S^+, it has a Laurent expansion of the form

$$\Omega(z) = \alpha_0 + \alpha_1 z + \alpha_2 z^2 + \cdots \tag{4.33}$$

so that

$$\Omega'(z) = \alpha_1 + 2\alpha_2 z + 3\alpha_3 z^2 + \cdots.$$

Also

$$\int_\Gamma \frac{t\,\overline{\Omega'(t)}\,dt}{t-z} = \int_\Gamma \frac{t\,\overline{\Omega'(a^2/\bar{t})}\,dt}{t-z} \quad (z \in S^+), \tag{4.34}$$

however $\overline{\Omega'(a^2/\bar{t})}$ is the boundary value of the function

$$\overline{\Omega'(a^2/\bar{\zeta})} = \bar{\alpha}_1 + 2\bar{\alpha}_2 a^2 \zeta^{-1} + 3\bar{\alpha}_3 a^4 \zeta^{-2} + \cdots$$

on $|\zeta| = a$ which is holomorphic in S^- $(|\zeta| > a)$. Hence as the integral in (4.34) is holomorphic for $\zeta \in S^-$ we can replace the integral around Γ by an integral around $|\zeta| = R$ as $R \to \infty$ giving

$$\begin{aligned}\int_\Gamma \frac{t\,\overline{\Omega'(a^2/\bar{t})}\,dt}{t-z} &= \int_{|\zeta|=R} \frac{\zeta\,\overline{\Omega'(a^2/\bar{\zeta})}\,d\zeta}{\zeta - z} \\ &= \int_{|\zeta|=R} (\bar{\alpha}_1 + 2\bar{\alpha}_2 a^2 \zeta^{-1} + \cdots)(1 - z\zeta^{-1})^{-1}\,d\zeta \\ &= 2\pi i(\bar{\alpha}_1 z + 2\bar{\alpha}_2 a^2).\end{aligned} \tag{4.35}$$

Thus from (4.31) $\Omega(z)$ satisfies the equation

$$\gamma\,\Omega(z) + \bar{\alpha}_1 z + 2\bar{\alpha}_2 a^2 = \frac{1}{2\pi i}\int_\Gamma \frac{F(t)\,dt}{t-z} \tag{4.36}$$

where α_1, α_2 are related to $\Omega(z)$ by (4.33). These constants are most easily evaluated by taking the Laurent expansion of (4.36) about the origin. Thus

$$\gamma(\alpha_0 + \alpha_1 z + \alpha_2 z^2 + \cdots) + \bar{\alpha}_1 z + 2\bar{\alpha}_2 a^2 = \frac{1}{2\pi i}\int_\Gamma \frac{F(t)}{t}\left(1 - \frac{z}{t}\right)^{-1} dt$$

and hence

$$\begin{aligned}\gamma\alpha_0 + 2\bar{\alpha}_2 a^2 &= \frac{1}{2\pi i}\int_\Gamma \frac{F(t)\,dt}{t} \\ \gamma\alpha_1 + \bar{\alpha}_1 &= \frac{1}{2\pi i}\int_\Gamma \frac{F(t)\,dt}{t^2} \\ \gamma\alpha_2 &= \frac{1}{2\pi i}\int_\Gamma \frac{F(t)\,dt}{t^3}.\end{aligned} \tag{4.37}$$

For the displacement problem where $\gamma = -\kappa$ equations (4.37) uniquely define α_0, α_1, α_2. For the stress problem where $\gamma = 1$ the second equation of (4.37) implies

$$\frac{1}{2\pi i}\int_\Gamma \frac{F(t)\,dt}{t^2} = -\frac{1}{2\pi a^2 i}\int_\Gamma F(t)\,d\bar{t} \quad \text{on } t\bar{t} = a^2$$

is real. On comparing with (4.7) this will be seen to be the necessary condition that the resultant moment applied to the disc be zero. Corresponding to this only the real part of α_1 is specified, the imaginary part merely contributing a rigid-body rotation to the solution.

Having found $\Omega(z)$ from (4.36) and (4.37), we may use Cauchy's formula (Section 1.5) to determine $\omega(z)$ in terms of its boundary value which is known from (4.28). Thus

$$\omega(z) = \frac{1}{2\pi i}\int_\Gamma \frac{\omega(t)\,dt}{t-z} = \frac{1}{2\pi i}\int_\Gamma \frac{\overline{F(t)} - \gamma\,\overline{\Omega(t)} - \bar{t}\,\Omega'(t)\,dt}{t-z}, \quad \text{for } z \in S^+$$

and again the integrals involving $\Omega(t)$ may be simply evaluated by residue theory

$$\frac{1}{2\pi i}\int_\Gamma \frac{\overline{\Omega(t)}\,dt}{t-z} = \overline{\Omega(0)} = \bar{\alpha}_0$$

$$\frac{1}{2\pi i}\int_\Gamma \frac{\bar{t}\,\Omega'(t)\,dt}{t-z} = \frac{1}{2\pi i}\int_\Gamma \frac{a^2\,\Omega'(t)\,dt}{t(t-z)}$$

$$= \frac{a^2}{z}\{\Omega'(z) - \Omega'(0)\}.$$

Hence

$$\omega(z) = \frac{1}{2\pi i}\int_\Gamma \frac{\overline{F(t)}\,dt}{t-z} - \gamma\bar{\alpha}_0 - a^2 z^{-1}\{\Omega'(z) - \alpha_1\} \tag{4.38}$$

and it may be confirmed that $\omega(0) = 0$ which was assumed in (4.30). For the stress problem as the complex potentials are only determined to within arbitrary constants the constant terms $2a^2\alpha_2$ in (4.36) and $\gamma\bar{\alpha}_0$ in (4.38) may be omitted, slightly simplifying the results.

It is clear that exactly the same method may be applied to the stress or displacement boundary value problems for an infinite medium containing a circular hole where, in the boundary condition (4.28), $F(t)$ is defined by (4.4) or (4.8). We shall however omit the details of this calculation as it is performed for a more general situation in Section 5.3. Since the solutions to these problems merely involve contour integrals around the circle Γ this method is very convenient for applications.† It is not easily generalized to more awkward boundary value problems such as the mixed problem, as it leads to singular integral equations. The mixed problem has been solved using this method by Sherman[53] and Sherman's work is briefly discussed by Mikhlin,[39] Chapter IX.

† As was mentioned in Section 3.11 an equivalent method based on the properties of Cauchy integrals on straight lines yields an equally useful solution for the stress or displacement boundary value problems for a half-plane. This approach has been given by Muskhelishvili,[43] Chapter 16.

We now consider a simple illustration of this method. Other illustrations are given in Section 5.3.

1. *Disc under equal and opposite point forces*

As shown in Figure 4.1, we consider a disc under the stress distribution

$$\tau_{rr} + \mathrm{i}\tau_{r\theta} = \frac{F\mathrm{e}^{-\mathrm{i}\alpha}}{a}\,\delta(\theta - \alpha) + \frac{F\mathrm{e}^{\mathrm{i}\alpha}}{a}\,\delta(\theta - \pi + \alpha) \quad \text{on } t = a\mathrm{e}^{\mathrm{i}\theta} \tag{4.39}$$

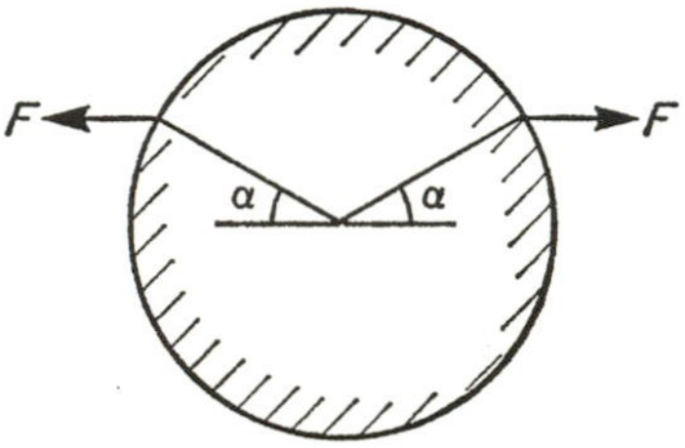

Figure 4.1

where $\delta(x)$ is the Dirac delta function. Then from (4.6) since $F(t)$ is proportional to the resultant force over an arc of the boundary

$$F(a\mathrm{e}^{\mathrm{i}\theta})\begin{cases} = 0 & (0 \leqslant \theta < \alpha) \\ = \mathrm{i}F & (\alpha \leqslant \theta \leqslant \pi - \alpha) \\ = 0 & (\pi - \alpha < \theta \leqslant 2\pi) \end{cases} \tag{4.40}$$

and on substituting in (4.36) and (4.37) we find

$$\Omega(z) = \frac{F}{2\pi}\int_{\theta=\alpha}^{\theta=\pi-\alpha} \frac{\mathrm{d}t}{t - z} - \bar{\alpha}_1 z$$

where

$$\bar{\alpha}_1 = \frac{F}{4\pi}\int_{\theta=\alpha}^{\theta=\pi-\alpha} \frac{\mathrm{d}t}{t^2}.$$

Thus

$$\Omega(z) = \frac{F}{2\pi}\log\left(\frac{z + a\mathrm{e}^{-\mathrm{i}\alpha}}{z - a\mathrm{e}^{\mathrm{i}\alpha}}\right) - \frac{zF\cos\alpha}{2\pi a} \tag{4.41}$$

and similarly from (4.38)

$$\omega(z) = -\frac{F}{2\pi}\log\left(\frac{z + a\mathrm{e}^{-\mathrm{i}\alpha}}{z - a\mathrm{e}^{\mathrm{i}\alpha}}\right) + \frac{F}{2\pi}\left(\frac{a\mathrm{e}^{\mathrm{i}\alpha}}{z + a\mathrm{e}^{-\mathrm{i}\alpha}} + \frac{a\mathrm{e}^{-\mathrm{i}\alpha}}{z - a\mathrm{e}^{\mathrm{i}\alpha}}\right). \tag{4.42}$$

The solution for diametrically opposed forces is given when $\alpha = 0$. It should be noted that the series solution of Section 4.2 would not immediately yield the above solution in a closed form.

4.4 Stress and displacement continuations

Let us suppose the region occupied by the elastic medium is either the interior or the exterior of the circle $|z| = a$ and denote this region by S^+, and its complement by S^-. In Section 3.3 we defined the stress continuation across the boundary $y = 0$ of the half-plane by supposing the applied stress $\tau_{yy} - i\tau_{xy}$ was zero on the boundary $y = 0$. Similarly the displacement continuation resulted from assuming $D = 0$ on $y = 0$. We shall employ the same methods of definition in this section. It was shown in Sections 2.11 and 4.1 that the stress or displacement boundary value problems have boundary conditions of the form

$$\gamma\, \Omega(t) + t\, \overline{\Omega'(t)} + \overline{\omega(t)} = F(t) \quad \text{on } t = a e^{i\theta} \tag{4.43}$$

where for the displacement problem $\gamma = -\kappa$ and $F(t) = -2\mu\, D(t)$ where $D(t)$ is the specified displacement on the boundary. Similarly in the stress problem $\gamma = 1$ and $F(t)$ is proportional to the resultant force over an arc of the boundary, in fact for the circular boundary $t\bar{t} = a^2$

$$F(t) = i \int_{\phi=0}^{\phi=\theta} (\tau_{rr} + i\tau_{r\theta}) e^{i\phi} a\, d\phi \quad (t = a e^{i\theta})$$

the lower limit in the integral being at our disposal and the direction of integration being chosen so that S^+ lies to the left. In Section 4.1 it was found expedient to use the relation (4.3) and to form the boundary conditions in terms of the single-valued potentials $\Omega_0(z)$, $\omega_0(z)$; it is convenient in this instance to deal with the complete potentials $\Omega(z)$, $\omega(z)$.

In order to define the continuations we assume either $\tau_{rr} + i\tau_{r\theta} = 0$ on the boundary $t\bar{t} = a^2$ or $D(t) = 0$. In either case we are led to the boundary condition

$$\gamma\, \Omega^+(t) + t\, \overline{\Omega'^+(t)} + \overline{\omega^+(t)} = 0 \quad \text{on } t\bar{t} = a^2$$

where the suffix $^+$ indicates the limiting value attained by a function defined in S^+ on the boundary $t\bar{t} = a^2$. This boundary condition may be expressed in terms of the associated functions† $\overline{\Omega(a^2/\bar{z})}$, $\overline{\omega(a^2/\bar{z})}$ as

$$\Omega^+(t) = -\frac{1}{\gamma}\{t\, \overline{\Omega'^-(a^2/\bar{t})} + \overline{\omega^-(a^2/\bar{t})}\} \tag{4.44}$$

† See Section 1.3.

on writing $t = a^2/\bar{t}$ and noting that

$$\overline{\omega^+(t)} = \lim_{\substack{z\to t\\ z\in S^+}} \overline{\omega(z)} = \lim_{\substack{z\to t\\ z\in S^-}} \overline{\omega(a^2/\bar{z})} = \overline{\omega^-(a^2/\bar{t})}$$

a similar relation holding for $\overline{\Omega'(t)}$. Now as the potentials $\Omega(z)$, $\omega(z)$ are holomorphic in S^+ and the points z, $a^2/\bar{z}$ are inverse points with respect to the circle $|z| = a$, so that for $z \in S^-$, $a^2/\bar{z} \in S^+$ and vice versa, the associated potentials $\overline{\Omega'(a^2/\bar{z})}$, $\overline{\omega(a^2/\bar{z})}$ are holomorphic for $z \in S^-$ (except for an isolated singularity at $z = 0$ if S^+ is the region $|z| > a$).† Thus (4.44) expresses the fact that the holomorphic functions $\Omega(z)$ for $z \in S^+$ and $-\{z\,\overline{\Omega'(a^2/\bar{z})} + \overline{\omega(a^2/\bar{z})}\}\gamma^{-1}$ for $z \in S^-$ are continuous across $|z| = a$. Hence it is natural to extend the definition of $\Omega(z)$ from S^+ into S^- by putting

$$\Omega(z) = -\frac{1}{\gamma}\{z\,\overline{\Omega'(a^2/\bar{z})} + \overline{\omega(a^2/\bar{z})}\} \quad \text{for } z \in S^- \tag{4.45}$$

and we see that $\Omega(z)$ is holomorphic in S^+ and S^- (except at $z = 0$ when S^+ is the region $|z| \geqslant a$) and is continuous across the arcs of $|z| = a$ on which $F(t) = 0$ in (4.43). Since the stress continuation is most convenient in applications we shall examine it in detail.

As the stress continuation occurs when $\gamma = 1$, $\Omega(z)$ is continued into S^- by the definition

$$\Omega(z) = -\{z\,\overline{\Omega'(a^2/\bar{z})} + \overline{\omega(a^2/\bar{z})}\} \quad (z \in S^-) \tag{4.46}$$

and on rearranging we find

$$\omega(z) = -\overline{\Omega(a^2/\bar{z})} - (a^2/z)\,\Omega'(z). \tag{4.47}$$

Again it should be emphasized that (4.47) could be regarded as a convenient definition of $\omega(z)$, $z \in S^+$, in terms of $\Omega(z)$ in S^+ and S^- where the values of $\Omega(z)$ in these two regions are quite independent. Further, since $\Omega(z)$, $\omega(z)$ are holomorphic in S^+, the behaviour of $\Omega(z)$ in S^- is determined by (4.46). In particular if S^+ is the region $|z| < a$ then $\Omega(z)$ is holomorphic in the whole of S^- $(|z| > a)$ and such that

$$\Omega'(z) = -\overline{\Omega'(0)} + O\left(\frac{1}{z^2}\right), \quad \text{for large } |z|. \tag{4.48}$$

Similarly, on differentiating (4.46), if S^+ is the region $|z| > a$ then $\Omega(z)$ is holomorphic in S^- except at the origin where $\Omega'(z)$ has a pole of the form

$$\Omega'(z) = (C - \mathrm{i}D)\frac{a^2}{z^2} + \frac{\kappa(X + \mathrm{i}Y)}{2\pi(1+\kappa)}\frac{1}{z} + \text{constant}, \quad \text{for small } |z| \tag{4.49}$$

† See Section 1.3.

since, from Section 2.10, $\Omega(z)$, $\omega(z)$ behave as

$$\begin{aligned} \Omega'(z) &= A + \mathrm{i}B - \frac{(X + \mathrm{i}Y)}{2\pi(1 + \kappa)}\frac{1}{z} + O\left(\frac{1}{z^2}\right) \\ \omega'(z) &= C + \mathrm{i}D + \frac{\kappa(X - \mathrm{i}Y)}{2\pi(1 + \kappa)}\frac{1}{z} + O\left(\frac{1}{z^2}\right) \end{aligned} \tag{4.50}$$

for large $|z|$.

As the behaviour of $\Omega(z)$ is known throughout the whole plane we may eliminate $\omega(z)$ from the stress and displacement relations (4.1) by substituting from (4.47). Thus,

$$\begin{aligned} 2\mu(u_r + \mathrm{i}u_\theta) &= 2\mu D\mathrm{e}^{-\mathrm{i}\theta} = \mathrm{e}^{-\mathrm{i}\theta}\left\{\kappa\,\Omega(z) + \Omega\left(\frac{a^2}{\bar{z}}\right) + \left(\frac{a^2}{\bar{z}} - z\right)\overline{\Omega'(z)}\right\} \\ \tau_{rr} + \tau_{\theta\theta} &= 2\{\Omega'(z) + \overline{\Omega'(z)}\} \\ \tau_{rr} + \mathrm{i}\tau_{r\theta} &= \Omega'(z) - \frac{a^2}{z\bar{z}}\,\Omega'\left(\frac{a^2}{\bar{z}}\right) + \left(1 - \frac{a^2}{z\bar{z}}\right)\{\overline{\Omega'(z)} - \bar{z}\,\overline{\Omega''(z)}\} \end{aligned} \tag{4.51}$$

and we note that on the boundary $t\bar{t} = a^2$

$$\begin{aligned} 2\mu D &= \kappa\,\Omega^+(t) + \Omega^-(t) \\ \tau_{rr} + \mathrm{i}\tau_{r\theta} &= \Omega'^+(t) - \Omega'^-(t). \end{aligned} \tag{4.52}$$

It will be seen that the stress continuation is such that $\Omega'(z)$ is continuous across those parts of $|z| = a$ on which $\tau_{rr} + \mathrm{i}\tau_{r\theta} = 0$.

Similarly for the displacement continuation from (4.45) where $\gamma = -\kappa$, the formulae corresponding to (4.46) and (4.47) are

$$\Omega(z) = \frac{1}{\kappa}\left\{z\,\overline{\Omega'\left(\frac{a^2}{\bar{z}}\right)} + \overline{\omega\left(\frac{a^2}{\bar{z}}\right)}\right\} \quad (z \in S^-) \tag{4.53}$$

and

$$\omega(z) = \kappa\,\overline{\Omega\left(\frac{a^2}{\bar{z}}\right)} - \frac{a^2}{z}\,\Omega'(z) \qquad (z \in S^+).$$

Again this last relation could be regarded as a useful substitution for $\omega(z)$ which reduces displacement boundary value problems to simple Cauchy boundary conditions for regions with circular boundaries. Formulae (4.46), (4.47) and (4.53) should be contrasted with the corresponding definitions in Section 3.3.

We shall now examine the applications of this method to the stress and displacement boundary value problems for a circular disc, an infinite medium containing a circular hole, and an annular region.

4.5 Solutions to the stress and displacement problems by continuation

1. *Stress problem for a circular disc*

Suppose the stress $\tau_{rr} + i\tau_{r\theta} = f(t)$ is given on the boundary $t = ae^{i\theta}$ of the disc $|z| \leqslant a$. On using the stress continuation from (4.52) the complex potential $\Omega'(z)$ is holomorphic in S^+ and S^- and such that on $|z| = a$

$$\Omega'^+(t) - \Omega'^-(t) = f(t). \quad (t = ae^{i\theta}).$$

Hence from the Plemelj formulae of Section 1.6

$$\Omega'(z) = \frac{1}{2\pi i}\int_\Gamma \frac{f(t)\,dt}{t - z} + \phi(z) \tag{4.54}$$

where Γ is the contour $|t| = a$ taken in an anticlockwise direction and $\phi(z)$ is holomorphic in the whole plane except possibly at the origin and infinity. However $\Omega'(z)$ is bounded at $z = 0$ and $z = \infty$ and hence $\phi(z)$ is at most a constant. If we put $\phi(z) = \alpha$ then

$$\Omega'(\infty) = \alpha, \qquad \Omega'(0) = \frac{1}{2\pi i}\int_\Gamma \frac{f(t)\,dt}{t} + \alpha$$

and for a circular disc the stress continuation implies $\Omega'(z)$ satisfies (4.48) so that

$$\alpha + \bar{\alpha} = -\frac{1}{2\pi i}\int_\Gamma \frac{f(t)\,dt}{t}. \tag{4.55}$$

Further in (4.48) the coefficient of the term in $1/z$ is zero and hence from (4.54)

$$\int_\Gamma f(t)\,dt = 0. \tag{4.56}$$

As has been seen previously in Section 4.1, equation (4.56) and the condition that the right-hand side of (4.55) be real, merely express the necessary conditions that the applied stress distribution has zero resultant force and zero resultant moment respectively. The imaginary part of α contributes a rigid-body rotation to the solution and is taken to be zero.

As an example if the disc is compressed under a uniform constant pressure p then $f(t) = -p$ and from (4.54) and (4.55) we find

$$\Omega'(z)\begin{cases} = -\tfrac{1}{2}p & (|z| \leqslant a) \\ = \tfrac{1}{2}p & (|z| \geqslant a). \end{cases}$$

The stress and displacement field may be determined immediately from (4.51).

In a similar manner to the above, the solution to the displacement boundary value problem for the circular disc may be determined. Assuming the boundary condition is $D = f(t)$, on substituting the displacement continuation (4.53) into (4.1) the boundary conditions become

$$\Omega^+(t) - \Omega^-(t) = \frac{2\mu}{\kappa} f(t) \quad \text{on } t\bar{t} = a^2.$$

This is a simple Cauchy problem for the holomorphic function $\Omega(z)$ and the solution is

$$\Omega(z) = \frac{\mu}{\kappa\pi \mathrm{i}} \int_\Gamma \frac{f(t)\,\mathrm{d}t}{t - z} + \alpha_0 + \alpha_1 z$$

where

$$\kappa\bar{\alpha}_1 - \alpha_1 = \frac{\mu}{\kappa\pi \mathrm{i}} \int_\Gamma \frac{f(t)\,\mathrm{d}t}{t^2}$$

$$\kappa\bar{\alpha}_0 = \frac{2a^2\mu}{\kappa\pi \mathrm{i}} \int_\Gamma \frac{f(t)\,\mathrm{d}t}{t^3}$$

and we have supposed $\omega(0) = 0$. Compare with equations (4.37).

2. *Stress problem for a circular hole in an infinite plate*

In this case if we suppose the stress and rotation are known at infinity and that the stress distribution $\tau_{rr} + \mathrm{i}\tau_{r\theta} = f(t)$ acts over the hole $t = a\mathrm{e}^{\mathrm{i}\theta}$ then, on using the stress continuation (4.46), $\Omega'(z)$ is defined in S^+ and S^-, having a pole of order 2 at $z = 0$ from (4.49), behaves at infinity according to (4.50), where $A + \mathrm{i}B$, $C + \mathrm{i}D$ are known constants, and is such that on the hole

$$\Omega'^+(t) - \Omega'^-(t) = f(t) \quad (t = a\mathrm{e}^{\mathrm{i}\theta}).$$

Then from the Plemelj formulae of Section 1.6

$$\Omega'(z) = -\frac{1}{2\pi \mathrm{i}} \int_\Gamma \frac{f(t)\,\mathrm{d}t}{t - z} + \psi(z) \tag{4.57}$$

where Γ is the circle $|t| = a$ described in an anticlockwise manner and $\psi(z)$ is a function holomorphic in the whole plane except possibly at the origin and infinity. From the behaviour of $\Omega'(z)$ at these points it is apparent that $\psi(z)$ can only have the form

$$\psi(z) = \alpha_0 + \frac{\alpha_1}{z} + \frac{\alpha_2}{z^2} \tag{4.58}$$

and on comparing (4.57) and (4.58) with (4.49) and (4.50) we find

$$\alpha_0 = A + \mathrm{i}B, \qquad \alpha_1 = \frac{\kappa(X + \mathrm{i}Y)}{2\pi(1 + \kappa)}, \qquad \alpha_2 = (C - \mathrm{i}D)a^2$$

where

$$X + \mathrm{i}Y = \mathrm{i}\int_\Gamma f(t)\,\mathrm{d}t$$

and we note again that Γ is described in an anticlockwise manner. Thus the general solution is

$$\Omega'(z) = -\frac{1}{2\pi\mathrm{i}}\int_\Gamma \frac{f(t)\,\mathrm{d}t}{t - z} + A + \mathrm{i}B + \frac{\kappa(X + \mathrm{i}Y)}{2\pi(1 + \kappa)}\frac{1}{z} + (C - \mathrm{i}D)\frac{a^2}{z^2}. \tag{4.59}$$

Particular examples may now be evaluated. If we suppose the stress distribution

$$\tau_{rr} + \mathrm{i}\tau_{r\theta} = -\frac{X + \mathrm{i}Y}{2\pi t}, \quad \text{on } t = a\mathrm{e}^{\mathrm{i}\theta}$$

acts over the hole with zero stress and rotation at infinity, then

$$\Omega'(z) = \frac{(X + \mathrm{i}Y)}{4\pi^2\mathrm{i}}\int_\Gamma \frac{\mathrm{d}t}{t(t - z)} + \frac{\kappa(X + \mathrm{i}Y)}{2\pi(1 + \kappa)}\frac{1}{z}.$$

The integral may be simply evaluated and gives

$$\int_\Gamma \frac{\mathrm{d}t}{t(t - z)}\begin{cases} = 0 & \text{for } |z| < a \\ = \dfrac{2\pi\mathrm{i}}{z} & \text{for } |z| > a \end{cases}$$

and thus

$$\Omega'(z)\begin{cases} = \dfrac{\kappa(X + \mathrm{i}Y)}{2\pi(1 + \kappa)}\dfrac{1}{z} & \text{for } |z| < a \\ = -\dfrac{(X + \mathrm{i}Y)}{2\pi(1 + \kappa)}\dfrac{1}{z} & \text{for } |z| > a. \end{cases}$$

This result should be compared with the series solution as given in (4.27). The displacement problem may be solved similarly.

3. *Stress concentrations*

So far we have been principally concerned with the various methods of solution and not deductions from these solutions. If we examine the case of an unstressed circular hole in an infinite plate under a uniform tension

$\tau_{xx} = T$ at infinity then from (2.61) and (4.59) we find

$$\Omega'(z) = \frac{T}{4}\left(1 - \frac{2a^2}{z^2}\right).$$

Hence on the hole the hoop stress $\tau_{\theta\theta}$ is given by

$$\tau_{rr} + \tau_{\theta\theta} = \tau_{\theta\theta} = 4\,\mathrm{Re}\{\Omega'(z)\} = T(1 - 2\cos 2\theta)$$

since $\tau_{rr} = 0$ on $|z| = a$. The hoop stress therefore attains a maximum value of $3T$ when $\theta = \pm\frac{1}{2}\pi$, that is at the ends of the diameter perpendicular to the direction of the tension at infinity, and a minimum value of $-T$ at $\theta = 0, \pi$. Thus the effect of a hole in a body is to change the existing stress field so that the stress level is appreciably raised in certain regions near the hole. The presence of such stress concentrations is of great practical importance since they may give rise to plastic deformation and fatigue cracking in metal components.

4. *A circular disc with internal point forces and moments*

As a fourth illustration of the method of continuation we consider the case of a circular disc $|z| \leqslant a$ under the action of a finite number of point forces and moments acting at internal points of the disc and suppose the edge of the disc to be unstressed. If the potentials $\Omega_0(z)$, $\omega_0(z)$ represent the system of point forces and moments when acting on a body occupying the whole plane then $\Omega_0(z)$, $\omega_0(z)$ are sums of terms of the form given in equation (3.6). It should be noted that $\Omega_0(z)$, $\omega_0(z)$ have a finite number of isolated singularities in $|z| < a$, and are holomorphic elsewhere. In particular the stress on the circle $|z| = a$ in the infinite plane is, from (4.1),

$$\tau_{rr} + i\tau_{r\theta} = \Omega_0'(t) + \overline{\Omega_0'(t)} - \overline{t\,\Omega_0''(t)} - t^{-1}\overline{t\,\omega_0'(t)}. \qquad (4.60)$$

We must also assume that the applied system of point forces and moments is in equilibrium so that from (3.6) for large $|z|$

$$\Omega_0(z) = O(z^{-1}), \quad \omega_0(z) = Az^{-1} + O(z^{-2}) \qquad (4.61)$$

where A is real.

If we add the stress system corresponding to potentials $\Omega_1(z)$, $\omega_1(z)$ so that the combined potentials $\Omega_1(z) + \Omega_0(z)$, $\omega_1(z) + \omega_0(z)$ have the correct singularities in $|z| < a$ and give rise to zero applied stress on $|z| = a$, then $\Omega_1(z)$, $\omega_1(z)$ must be holomorphic in $|z| < a$ (giving rise to no further singularities) and the additional stress system must remove the stress distribution (4.60) on the boundary. If we now use the stress continuation (4.46) for the potentials $\Omega_1(z)$, $\omega_1(z)$ so that $\Omega_1(z)$ is defined and holomorphic in $|z| < a$ and $|z| > a$ and such that (4.48) holds, then from

(4.52) the additional stress system satisfies the relations

$$\begin{aligned}\tau_{rr} + i\tau_{r\theta} &= \Omega_1'^{+}(t) - \Omega_1'^{-}(t)\\ &= -\{\Omega_0'(t) + \overline{\Omega_0'(t)} - \bar{t}\,\overline{\Omega_0''(t)} - t^{-1}\bar{t}\,\overline{\omega_0'(t)}\}\end{aligned} \tag{4.62}$$

for an unstressed boundary.

Following the method of the first illustration these boundary conditions may be satisfied by representing $\Omega_1(z)$ in terms of a Cauchy integral as in (4.54). However it is more direct to observe that (4.62) may be written in the form

$$\begin{aligned}\Omega_1'^{+}(t) - \Omega_1'^{-}(t) = -\Omega_0'(t) &- \overline{\Omega_0'(a^2/\bar{t})}\\ &+ t^{-1}a^2\,\overline{\Omega_0''(a^2/\bar{t})} + t^{-2}a^2\,\overline{\omega_0'(a^2/\bar{t})}\end{aligned}$$

and hence the boundary conditions may be satisfied by putting

$$\begin{aligned}\Omega_1'(z) &= -\overline{\Omega_0'(a^2/\bar{z})} + z^{-1}a^2\,\overline{\Omega_0''(a^2/\bar{z})} + z^{-2}a^2\,\overline{\omega_0'(a^2/\bar{z})} + \phi(z) && (|z| < a)\\ \Omega_1'(z) &= \Omega_0'(z) + \phi(z) && (|z| > a)\end{aligned} \tag{4.63}$$

since the functions on the right-hand side are holomorphic in the regions indicated. The additional function $\phi(z)$ is holomorphic in the whole plane except possibly at the origin and infinity. However since $\Omega_0(z)$, $\omega_0(z)$ satisfy (4.61), $\phi(z)$ is at most a complex constant, $\phi(z) = \alpha$. This constant may be determined by noting that $\Omega_1(z)$ satisfies (4.48) so that

$$\alpha + \bar{\alpha} = A/a^2$$

where A has been defined in (4.61).

Finally on integrating equations (4.63) we find

$$\begin{aligned}\Omega_1(z) &= -z\,\overline{\Omega_0'(a^2/\bar{z})} - \overline{\omega_0(a^2/\bar{z})} + \tfrac{1}{2}Aza^{-2} && (|z| < a)\\ \Omega_1(z) &= \Omega_0(z) + \tfrac{1}{2}Aza^{-2} && (|z| > a)\end{aligned} \tag{4.64}$$

to within arbitrary constants corresponding to a rigid-body motion. On using (4.47) $\omega_1(z)$ may be found and hence the complete potentials $\Omega_0(z) + \Omega_1(z)$, $\omega_0(z) + \omega_1(z)$ determined. Particular examples of this general solution are left as an exercise. It is clear that the same method may be applied in the case of a circular hole in an infinite medium, and for example if the boundary of the hole or the disc is clamped so that $D = 0$.

5. *Stress and displacement problems for an annulus*

The stress and displacement boundary value problems for an annular region have been solved in Section 4.2 using the series solution. In this section we briefly describe how the stress and displacement continuations derived in the last section may be used in these problems. This method has been employed by Milne-Thomson,[41] Section 5.5 and Buchwald has corrected and applied Milne-Thomson's work to some problems involving doubly connected regions.†

We denote the annular region $a < |z| < b$ by A and the annuli $a^2b^{-1} < |z| < a$, $b < |z| < b^2a^{-1}$ by A^- and A^+ respectively, see Figure 4.2. Now if we suppose either the normal stress $\tau_{rr} + i\tau_{r\theta}$ or the

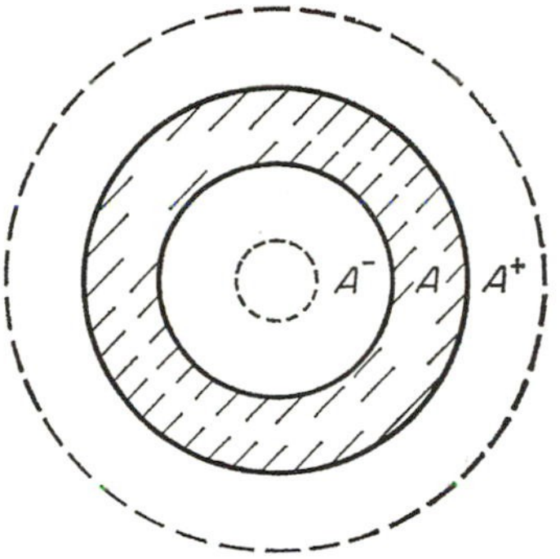

Figure 4.2

displacements $u + iv$ are specified on the boundaries $|z| = a$, $|z| = b$ of A, the boundary conditions become, from (4.11)

$$\begin{aligned} \gamma_1\, \Omega(t) + t\,\overline{\Omega'(t)} + \overline{\omega(t)} &= F_1(t) \quad \text{on } t = ae^{i\theta} \\ \gamma_2\, \Omega(t) + t\,\overline{\Omega'(t)} + \overline{\omega(t)} &= F_2(t) \quad \text{on } t = be^{i\theta} \end{aligned} \tag{4.65}$$

where $\Omega(z)$, $\omega(z)$ are holomorphic and single valued in A, and $F_1(t)$, $F_2(t)$ are single-valued functions given by (4.4) or (4.8). If we now use the continuation (4.45) across each boundary we find $\Omega(z)$ is extended from A into the annuli A^-, A^+ by the definitions

$$\begin{aligned} \Omega(z) &= -\frac{1}{\gamma_1}\left\{z\,\overline{\Omega'\left(\frac{a^2}{\bar z}\right)} + \overline{\omega\left(\frac{a^2}{\bar z}\right)}\right\} \quad \text{for } z \in A^- \\ \Omega(z) &= -\frac{1}{\gamma_2}\left\{z\,\overline{\Omega'\left(\frac{b^2}{\bar z}\right)} + \overline{\omega\left(\frac{b^2}{\bar z}\right)}\right\} \quad \text{for } z \in A^+. \end{aligned} \tag{4.66}$$

$\Omega(z)$ is thus holomorphic in the three regions A^-, A, A^+.

† See for example Buchwald,[3] Buchwald and Davies,[5] and also Tiffen and Semple.[65]

If we invert these continuations we find

$$\begin{aligned}\omega(z) &= -\gamma_1\,\overline{\Omega\left(\frac{a^2}{\bar{z}}\right)} - \frac{a^2}{z}\,\Omega'(z) \quad (z\in A)\\ \omega(z) &= -\gamma_2\,\overline{\Omega\left(\frac{b^2}{\bar{z}}\right)} - \frac{b^2}{z}\,\Omega'(z) \quad (z\in A)\end{aligned} \tag{4.67}$$

and hence $\Omega(z)$ must satisfy the *compatibility identity*

$$\gamma_2\,\overline{\Omega\left(\frac{b^2}{\bar{z}}\right)} - \gamma_1\,\overline{\Omega\left(\frac{a^2}{\bar{z}}\right)} + \frac{(b^2-a^2)}{z}\,\Omega'(z) = 0 \tag{4.68}$$

for all $z \in A$.

On substituting equations (4.66) into the boundary conditions (4.65) we find

$$\begin{aligned}\Omega(t) - \Omega^-(t) &= F_1(t)/\gamma_1 \quad \text{on } t = a\mathrm{e}^{\mathrm{i}\theta}\\ \Omega(t) - \Omega^+(t) &= F_2(t)/\gamma_2 \quad \text{on } t = b\mathrm{e}^{\mathrm{i}\theta}\end{aligned} \tag{4.69}$$

where $\Omega^-(t)$ and $\Omega^+(t)$ denote the limits on $|z| = a$ and $|z| = b$ of $\Omega(z)$ in A^- and A^+ respectively.

Although Milne-Thomson[41] has expressed the solution to (4.69) in terms of Cauchy integrals it is more direct to use Buchwald's approach. Since $\Omega(z)$ is holomorphic and single valued in A^-, A, A^+ it may be represented by the Laurent series (Section 1.3)

$$\Omega(z)\begin{cases} = \sum\limits_{-\infty}^{\infty} A_n^-\,z^n & (z\in A^-)\\ = \sum\limits_{-\infty}^{\infty} A_n\,z^n & (z\in A)\\ = \sum\limits_{-\infty}^{\infty} A_n^+\,z^n & (z\in A^+)\end{cases} \tag{4.70}$$

in these three regions. Hence the boundary conditions (4.69) become

$$\begin{aligned}a^n(A_n - A_n^-) &= \frac{1}{2\pi\gamma_1}\int_{-\pi}^{\pi} F_1(a\mathrm{e}^{\mathrm{i}\theta})\mathrm{e}^{-\mathrm{i}n\theta}\,\mathrm{d}\theta\\ b^n(A_n - A_n^+) &= \frac{1}{2\pi\gamma_2}\int_{-\pi}^{\pi} F_2(b\mathrm{e}^{\mathrm{i}\theta})\mathrm{e}^{-\mathrm{i}n\theta}\,\mathrm{d}\theta\end{aligned} \tag{4.71}$$

and on substituting in the compatibility identity (4.68) we find

$$\gamma_2 b^{2n}A_n^+ - \gamma_1 a^{2n}A_n^- + (b^2-a^2)(2-n)\bar{A}_{2-n} = 0.$$

Hence the A_n satisfy the system of equations

$$\begin{aligned}&(\gamma_2 b^{2n} - \gamma_1 a^{2n})A_n + (b^2-a^2)(2-n)\bar{A}_{2-n}\\ &\qquad = \frac{1}{2\pi}\int_{-\pi}^{\pi}\{b^n F_2(b\mathrm{e}^{\mathrm{i}\theta}) - a^n F_1(a\mathrm{e}^{\mathrm{i}\theta})\}\mathrm{e}^{-\mathrm{i}n\theta}\,\mathrm{d}\theta.\end{aligned} \tag{4.72}$$

This is exactly the system of equations (4.14) which arises in the series solution to the problem. As a discussion of the equations (4.72) has been given in Section 4.2 we omit further comment except to remark that $\omega(z)$ may be immediately calculated from one of the relations (4.67) when the corresponding coefficients have been evaluated in (4.71). It will be noted that the amount of computation in the series solution and this method is exactly the same.

4.6 The mixed boundary value problem

Let us suppose that over a region L which is the union of a finite number of arcs of the circular boundary $|z| = a$ the displacement D is specified and over the remainder L' of $|z| = a$ the stress $\tau_{rr} + i\tau_{r\theta}$ is specified. Then these boundary conditions characterize the mixed boundary value problem for either the circular disc $|z| \leqslant a$ or an infinite medium with a circular hole $|z| \geqslant a$. As the stress boundary value problem has been solved for these regions in the earlier sections we can consider the slightly simpler boundary conditions

$$\begin{aligned} D = u + iv = f(t) \quad & \text{for } t = ae^{i\theta} \in L \\ \tau_{rr} + i\tau_{r\theta} = 0 \quad & \text{for } t = ae^{i\theta} \in L' \end{aligned} \tag{4.73}$$

without loss of generality. As in Section 3.7 these boundary conditions are most effectively handled by using the stress continuation defined in (4.46) and from (4.51) we find

$$\begin{aligned} 2\mu D &= \kappa\, \Omega(z) + \Omega\left(\frac{a^2}{\bar{z}}\right) + \left(\frac{a^2}{\bar{z}} - z\right)\overline{\Omega'(z)} \\ \tau_{rr} + i\tau_{r\theta} &= \Omega'(z) - \left(\frac{a^2}{z\bar{z}}\right)\Omega'\left(\frac{a^2}{\bar{z}}\right) + \left(1 - \frac{a^2}{z\bar{z}}\right)\{\overline{\Omega'(z)} - \bar{z}\,\overline{\Omega''(z)}\}. \end{aligned} \tag{4.74}$$

Since it is easier to avoid the presence of the logarithmic terms which occur in $\Omega(z)$ we express the boundary conditions in terms of $\Omega'(z)$. This is achieved by satisfying the displacement boundary condition of (4.73) in differentiated form. Now from (4.74)

$$2\mu \frac{\partial D}{\partial \theta} = it\{\kappa\, \Omega'(t) + \Omega'(a^2/\bar{t})\} \quad \text{on } t = ae^{i\theta}$$

and hence from (4.73) the boundary conditions become

$$\begin{aligned} \kappa\, \Omega'^{+}(t) + \Omega'^{-}(t) = 2\mu f'(t) \quad & (t \in L) \\ \Omega'^{+}(t) - \Omega'^{-}(t) = 0 \quad & (t \in L'). \end{aligned} \tag{4.75}$$

Thus the function $\Omega'(z)$ is holomorphic and single valued in the whole plane cut along the arcs L, over which it satisfies the condition (4.75), and has a known behaviour at the origin and infinity given by (4.48) or (4.49) and (4.50) according as the medium is a circular disc or the region outside a circular hole. Hence $\Omega'(z)$ satisfies a Hilbert problem of the type discussed in Section 1.6. If we take the sense of description of the arcs L in such a way as to keep the medium to the left and denote the end-points of the arcs by $a_1, b_1; a_2, b_2; \ldots; a_n, b_n$; (see Figure 4.3 which applies to

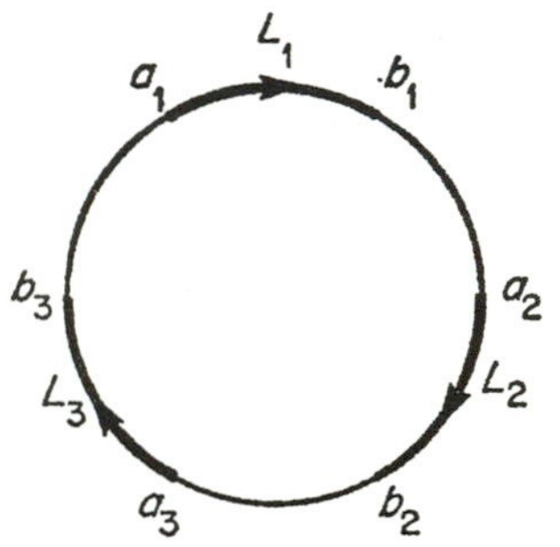

Figure 4.3

the case of a circular hole) then the basic Plemelj function is, from (1.28),

$$X(z) = \prod_{k=1}^{n} (z - a_k)^{-\gamma}(z - b_k)^{\gamma-1} \tag{4.76}$$

where

$$2\pi i\gamma = \log\left(\frac{-1}{\kappa}\right) = -\log(\kappa) + i\pi$$

so that

$$\gamma = \tfrac{1}{2} + i\beta, \qquad \beta = \frac{1}{2\pi}\log\kappa \tag{4.77}$$

and we select the branch of $X(z)$ so that $z^n X(z) \to 1$ as $|z| \to \infty$. Thus from Section 1.6 the general solution to the problem is

$$\Omega'(z) = \frac{\mu X(z)}{\kappa\pi i}\int_L \frac{f'(t)\,dt}{X^+(t)(t-z)} + \psi(z)X(z) \tag{4.78}$$

where $\psi(z)$ is an arbitrary function holomorphic in the whole plane except possibly at the origin and infinity where it must be chosen so that $\Omega'(z)$ has the correct behaviour.

For a circular disc $\Omega'(z)$ is holomorphic in the whole plane including the origin and infinity where it satisfies the condition (4.48). Hence $\psi(z)$ is at most a polynomial of degree n

$$\psi(z) = D_0 + D_1 z + \cdots + D_n z^n. \tag{4.79}$$

The condition (4.48) gives two relations for the constants D_i, one resulting from $\Omega'(\infty) = -\overline{\Omega'(0)}$, the other from the condition that the coefficient of the term in $1/z$ should be zero. The remaining constants are determined by additional physical assumptions similar to those used for the corresponding half-plane problem, namely that the total force acting over each arc of L is known, or that the relative depths of indentation of the punches are known, giving a further n equations. In all cases the equations resulting from (4.48) restrict the solution so that the resultant applied force and moment acting on the disc are zero.

For an infinite medium containing a circular hole $\Omega'(z)$ has a pole of order 2 at the origin of the form (4.49) and behaves at infinity according to (4.50) where the constants $A + \mathrm{i}B$, $C + \mathrm{i}D$ are given in terms of the known stress and rotation at infinity and $X + \mathrm{i}Y$ is the resultant force over the hole. Hence $\psi(z)$ must have the form

$$\psi(z) = \frac{D_{-2}}{z^2} + \frac{D_{-1}}{z} + D_0 + D_1 z + \cdots + D_n z^n. \tag{4.80}$$

It will be seen that the conditions at infinity (4.50) give two equations for the D_i, and the conditions at the origin (4.49) a further two equations. The remaining constants are determined on making additional physical assumptions of the sort described for the disc and the half-space.

As an illustration we consider the case where the displacement D is a constant over a single arc $L = (a\mathrm{e}^{\mathrm{i}\phi}, a\mathrm{e}^{-\mathrm{i}\phi})$ of the hole $|z| = a$, the remainder of the boundary being unstressed. We suppose the stress at infinity is specified and the resultant force $X + \mathrm{i}Y$ over the arc L is zero. These conditions may be regarded as equivalent to the deformation of a plate containing a circular hole under known stresses at infinity where the arc L of the hole is strongly reinforced so that L is only permitted a rigid-body translation (the case where L is permitted a small rotation is also simply evaluated). Supposing the stress at infinity is a uniform tension T at angle α to the x-axis, then from (2.61)

$$A = \tfrac{1}{4}T, \qquad C + \mathrm{i}D = -\tfrac{1}{2}T\mathrm{e}^{-2\mathrm{i}\alpha}.$$

Similarly if the rotation at infinity is zero, $B = 0$. Then from the solution given above, since $f'(t) = 0$, and $n = 1$

$$\Omega'(z) = \psi(z)(z - a\mathrm{e}^{\mathrm{i}\phi})^{-\gamma}(z - a\mathrm{e}^{-\mathrm{i}\phi})^{\gamma-1}$$

where

$$\psi(z) = \frac{D_{-2}}{z^2} + \frac{D_{-1}}{z} + D_0 + D_1 z.$$

In this case the constants follow directly from the behaviour of $\Omega'(z)$ at the origin and infinity. To evaluate this behaviour we require the forms of $X(z)$ near these points. By definition $z\,X(z) \to 1$ as $|z| \to \infty$, and hence for large $|z|$

$$X(z) = \frac{1}{z} + \frac{\gamma a e^{i\phi} + (1-\gamma)a e^{-i\phi}}{z^2} + O\left(\frac{1}{z^3}\right).$$

For small $|z|$, by a careful inspection of the arguments, we find (see Section 1.7)

$$X(z) = -a^{-1}e^{2\phi\beta}\{1 + a^{-1}z(\gamma e^{-i\phi} + (1-\gamma)e^{i\phi}) + \cdots\}$$

where β was defined in (4.77). Thus on comparing the expansions of $\Omega'(z)$ at infinity from (4.50) we find

$$D_1 = \tfrac{1}{4}T$$

$$D_0 + D_1(\gamma a e^{i\phi} + (1-\gamma)a e^{-i\phi}) = 0$$

since $X + iY = 0$, and similarly at the origin from (4.49)

$$2D_{-2} = Ta^3 e^{-2\phi\beta} e^{2i\alpha}$$

$$aD_{-1} + D_{-2}\{\gamma e^{-i\phi} + (1-\gamma)e^{i\phi}\} = 0.$$

Thus

$$\Omega'(z) = \tfrac{1}{4}T\left[z - a\{\gamma e^{i\phi} + (1-\gamma)e^{-i\phi}\} + 2a^2 e^{2i\alpha} e^{-2\phi\beta}\left\{\frac{a}{z^2} - \frac{\gamma e^{-i\phi} + (1-\gamma)e^{i\phi}}{z}\right\}\right](z - ae^{i\phi})^{-\gamma}(z - ae^{-i\phi})^{\gamma-1} \tag{4.81}$$

where $\gamma = \frac{1}{2} + i\beta = \frac{1}{2} + i(\log\kappa)/2\pi$. The stress field may now be calculated from (4.51). It will be seen from (4.74) and (4.81) that oscillating stress singularities of the form discussed in Section 3.7 occur at the ends of the arc L. A second example of this nature has been given by Muskhelishvili,[43] Section 123.

4.7 An infinite medium with cracks along arcs of a circle

We consider an infinite medium containing a set of line cracks which lie along the region L of the circle $|z| = a$, where L is the union of the set of arcs $a_1, b_1; a_2, b_2; \ldots; a_n, b_n$ described in an anticlockwise direction (see Figure 4.3). We suppose the body is deformed by known stresses at infinity and that in addition the cracks are opened by the application of internal pressures and shears so that

$$\begin{aligned}(\tau_{rr} + \mathrm{i}\tau_{r\theta})^+ &= f_1(t)\\ (\tau_{rr} + \mathrm{i}\tau_{r\theta})^- &= f_2(t)\end{aligned} \tag{4.82}$$

on the inner and outer edges of the cracks respectively. Across the remainder L' of $|z| = a$ the stress $\tau_{rr} + \mathrm{i}\tau_{r\theta}$ and the displacement D are continuous. To satisfy these boundary conditions we employ the analogue of the method of Section 3.10. Since D is continuous across region L' of $|z| = a$ then from (4.1)

$$\{\kappa\, \Omega(t) - t\, \overline{\Omega'(t)} - \overline{\omega(t)}\}^+ = \{\kappa\, \Omega(t) - t\, \overline{\Omega'(t)} - \overline{\omega(t)}\}^-$$

and as

$$\overline{\omega^+(t)} = \lim_{z\to t+} \overline{\omega(z)} = \lim_{z\to t-} \omega(a^2/\bar{z}) = \omega^-(a^2/\bar{t}),$$

$$\overline{\omega^-(t)} = \overline{\omega^+(a^2/\bar{t})}$$

and similarly for $\Omega'(z)$ we find

$$\kappa\, \Omega^+(t) + t\, \overline{\Omega'^+(a^2/\bar{t})} + \overline{\omega^+(a^2/\bar{t})} = \kappa\, \Omega'^-(t) + t\, \overline{\Omega'^-(a^2/\bar{t})} + \overline{\omega^-(a^2/\bar{t})}.$$

Thus the function

$$\kappa\, \Omega(z) + z\, \overline{\Omega'(a^2/\bar{z})} + \overline{\omega(a^2/\bar{z})} = \phi(z) \tag{4.83}$$

is holomorphic in the interior and exterior of the circle $|z| = a$ (except possibly at the origin) and is continuous across region L' of $|z| = a$. Similarly continuity of stress $\tau_{rr} + \mathrm{i}\tau_{r\theta}$ implies the function

$$\Omega(z) - z\, \overline{\Omega'(a^2/\bar{z})} - \overline{\omega(a^2/\bar{z})} = \theta(z) \tag{4.84}$$

is continuous across region L' of $|z| = a$. Hence $\theta(z)$, $\phi(z)$ are holomorphic in the whole plane cut along the arcs L except possibly at the origin. Since $\Omega'(z)$, $\omega'(z)$ behave as (4.50) for large $|z|$ and are both holomorphic at $z = 0$, it follows from (4.83) and (4.84) that for small $|z|$, $\theta'(z)$ has a pole of order 2 at the origin and is given by

$$\theta'(z) = (C - \mathrm{i}D)\frac{a^2}{z^2} + \frac{\kappa(X + \mathrm{i}Y)}{2\pi(1 + \kappa)}\frac{1}{z} + O(1) \tag{4.85}$$

and for large $|z|$ similarly

$$\theta'(z) = A + \mathrm{i}B - \overline{\Omega'(0)} - \frac{X + \mathrm{i}Y}{2\pi(1 + \kappa)}\frac{1}{z} + O\left(\frac{1}{z^2}\right). \tag{4.86}$$

We can now express $\Omega(z)$, $\omega(z)$ from equations (4.83), (4.84) in terms of

$\theta(z)$, $\phi(z)$ giving

$$(1 + \kappa)\,\Omega(z) = \theta(z) + \phi(z)$$
$$\omega(z) = \overline{\Omega\left(\frac{a^2}{\bar{z}}\right)} - \left(\frac{a^2}{z}\right)\Omega'(z) - \overline{\theta\left(\frac{a^2}{\bar{z}}\right)} \qquad (4.87)$$

and we note $\Omega(z)$ is holomorphic in the cut plane. The stresses and displacements from (4.1) may now be expressed in terms of $\theta(z)$ and $\phi(z)$, however as there is a simple linear relation between $\Omega(z)$, $\theta(z)$ and $\phi(z)$, it is sufficient to replace $\omega(z)$ by using equation (4.87). The relations (4.1) now become

$$2\mu D = \kappa\,\Omega(z) - \Omega\left(\frac{a^2}{\bar{z}}\right) + \left(\frac{a^2}{\bar{z}} - z\right)\overline{\Omega'(z)} + \theta\left(\frac{a^2}{\bar{z}}\right)$$
$$\tau_{rr} + \mathrm{i}\tau_{r\theta} = \Omega'(z) + \left(\frac{a^2}{z\bar{z}}\right)\Omega'\left(\frac{a^2}{\bar{z}}\right) + \left(1 - \frac{a^2}{z\bar{z}}\right)\left\{\overline{\Omega'(z)} - \bar{z}\,\overline{\Omega''(z)}\right\} \qquad (4.88)$$
$$- \left(\frac{a^2}{z\bar{z}}\right)\theta'\left(\frac{a^2}{\bar{z}}\right).$$

From (4.82) and (4.88) the stress boundary conditions on the cracks become

$$\left.\begin{aligned}\Omega'^{+}(t) + \Omega'^{-}(t) - \theta'^{-}(t) &= f_1(t)\\ \Omega'^{-}(t) + \Omega'^{+}(t) - \theta'^{+}(t) &= f_2(t)\end{aligned}\right\} \quad (t \in L)$$

and on subtracting and adding we find

$$\left.\begin{aligned}\theta'^{+}(t) - \theta'^{-}(t) &= f_1(t) - f_2(t)\\ \{2\Omega'(t) - \theta'(t)\}^{+} + \{2\Omega'(t) - \theta'(t)\}^{-} &= f_1(t) + f_2(t)\end{aligned}\right\} \quad (t \in L). \qquad (4.89)$$

Thus the functions $\theta'(z)$ and $2\Omega'(z) - \theta'(z)$ satisfy Hilbert problems for the set of arcs L, and behave at the origin and infinity according to (4.85), (4.86) and (4.50). The general solution of the first Hilbert problem of (4.89) is, from Section 1.6,

$$\theta'(z) = \frac{1}{2\pi\mathrm{i}}\int_L \frac{\{f_1(t) - f_2(t)\}\,\mathrm{d}t}{t - z} + \alpha_0 + \frac{\alpha_1}{z} + \frac{\alpha_2}{z^2}$$

where on comparison with (4.85) and (4.86) we find

$$\theta'(z) = \frac{1}{2\pi\mathrm{i}}\int_L \frac{\{f_1(t) - f_2(t)\}\,\mathrm{d}t}{t - z} + A + \mathrm{i}B - \overline{\Omega'(0)}$$
$$+ \frac{\kappa(X + \mathrm{i}Y)}{2\pi(1 + \kappa)}\frac{1}{z} + (C - \mathrm{i}D)\frac{a^2}{z^2}. \qquad (4.90)$$

Similarly the general solution to the second Hilbert problem of (4.89) is

$$2\,\Omega'(z) - \theta'(z) = \frac{X(z)}{2\pi\mathrm{i}}\int_L \frac{\{f_1(t) + f_2(t)\}\,\mathrm{d}t}{X^{+}(t)(t - z)}$$
$$+ \left(\frac{\beta_2}{z^2} + \frac{\beta_1}{z} + D_0 + D_1 z + \cdots + D_n z^n\right)X(z) \qquad (4.91)$$

where

$$X(z) = \prod_{k=1}^{n} (z - a_k)^{-1/2} (z - b_k)^{-1/2}$$

since $\gamma = \frac{1}{2}$ and we select the branch such that $z^n X(z) \to 1$ as $|z| \to \infty$. The $(n + 3)$ constants occurring in (4.91) are determined in part from comparison with the known behaviour of $2\Omega'(z) - \theta'(z)$ at the origin and infinity which yields four equations for the constants from (4.85), (4.86) and (4.50). As in Section 3.10, the other relations are found by ensuring the displacement field is non-dislocational and are

$$(1 + \kappa) \int_{a_k}^{b_k} \{\Omega'^{+}(t) - \Omega'^{-}(t)\}\, \mathrm{d}t = \int_{a_k}^{b_k} (f_1(t) - f_2(t))\, \mathrm{d}t$$
$$= \mathrm{i}(X_k + \mathrm{i}\,Y_k) \quad \text{for } k = 1, 2, \ldots, n \tag{4.92}$$

where $X_k + \mathrm{i}\,Y_k$ is the resultant force over the kth crack.† In fact one of these relations is degenerate as we have already taken into account the relation between $\Omega'(z)$ and the total resultant force $X + \mathrm{i}\,Y$.

As an illustration we consider a single arc crack on $|z| = a$, $|\theta| \leqslant \phi$ which is unstressed, $f_1(t) = f_2(t) = 0$, and under the action of a uniform tension T at infinity acting at an angle α to the x-axis. Then assuming the rotation at infinity is zero we have

$$A + \mathrm{i}B = \tfrac{1}{4}T, \qquad C + \mathrm{i}D = -\tfrac{1}{2}T\mathrm{e}^{-2\mathrm{i}\alpha}$$

so that

$$\theta'(z) = \tfrac{1}{4}T - \overline{\Omega'(0)} - \tfrac{1}{2}T\mathrm{e}^{2\mathrm{i}\alpha}a^2 z^{-2}$$

and

$$2\Omega'(z) - \theta'(z) = \left(\frac{\beta_2}{z^2} + \frac{\beta_1}{z} + D_0 + D_1 z\right)(z - a\mathrm{e}^{-\mathrm{i}\phi})^{-1/2}(z - a\mathrm{e}^{\mathrm{i}\phi})^{-1/2}. \tag{4.93}$$

To determine the remaining constants we require the expansions of $X(z)$ for large and small $|z|$, which have been given in Section 1.7, namely

$$(z - a\mathrm{e}^{-\mathrm{i}\phi})^{-1/2}(z - a\mathrm{e}^{\mathrm{i}\phi})^{-1/2} \begin{cases} = \dfrac{1}{z} + \dfrac{a\cos\phi}{z^2} + O\left(\dfrac{1}{z^3}\right) & \text{for large } |z| \\ = -\dfrac{1}{a}\left[1 + \dfrac{z\cos\phi}{a} + \dfrac{z^2(1 + 3\cos 2\phi)}{4a^2} + \cdots\right] & \text{for small } |z| \end{cases}$$

† These relations should be compared with (2.57) and (3.96).

(compare with Section 4.6 when $\gamma = \frac{1}{2}$). Hence for small $|z|$

$$\beta_2 = -1/2Te^{2i\alpha}a^3, \qquad a\beta_1 = -\beta_2 \cos\phi$$

for large $|z|$

$$D_1 = 1/4T + \overline{\Omega'(0)}, \qquad D_0 = -D_1 a \cos\phi.$$

We also see that on taking the limit as $z \to 0$ in (4.93)

$$2\Omega'(0) + \overline{\Omega'(0)}(1 - \cos\phi) = -\tfrac{1}{4}Te^{2i\alpha}\sin^2\phi + \tfrac{1}{4}T(1 + \cos\phi)$$

which determines $\Omega'(0)$. In the particular case of a semi-circular crack ($\phi = \frac{1}{2}\pi$) under a tension T along the x-axis ($\alpha = 0$) we find

$$\theta'(z) = \frac{T}{4}\left(1 - \frac{2a^2}{z^2}\right)$$

$$\Omega'(z) = \frac{T}{8}\left(1 + \frac{z}{(z^2 + a^2)^{1/2}}\right) - \frac{Ta^2}{4z^2}\left(1 + \frac{a}{(z^2 + a^2)^{1/2}}\right).$$

The stress field around the crack may now be evaluated.

The theory of this section may be used also to solve the problem when the displacements are specified on the edges of the cracks, or even more generally when the displacements and stresses are specified on the left and right edges respectively, but these are omitted from this text.

4.8 Frictionless boundary conditions

1. *Complete contact*

For many problems involving inclusions or bodies inserted into a hole in an elastic medium it is convenient to make the simplifying assumptions that the inclusion or insert is rigid and makes a frictionless contact with the surrounding elastic material. In these cases, assuming there is no separation between the insert and the material, it may be shown that the specification of the normal displacement and the shear stress on the boundary of insert are the most general boundary conditions that can occur. We consider the case of an elastic body occupying the region S^+ with a circular boundary $|z| = a$ and suppose

$$u_r = f(t), \qquad \tau_{r\theta} = g(t) \quad \text{on } t = ae^{i\theta} \tag{4.94}$$

where $f(t)$, $g(t)$ have real values. We denote the inverse region of S^+ in $|z| = a$ by S^- and the circle $|z| = a$ by Γ oriented so that S^+ lies to the left. From (4.1) we see

$$2i\tau_{r\theta} = z\,\Omega''(z) + z\bar{z}^{-1}\,\omega'(z) - \bar{z}\,\overline{\Omega''(z)} - \bar{z}z^{-1}\,\overline{\omega'(z)} \tag{4.95}$$

$$4\mu u_r = e^{-i\theta}\{\kappa\,\Omega(z) - z\,\overline{\Omega'(z)} - \overline{\omega(z)}\} + e^{i\theta}\{\kappa\,\overline{\Omega(z)} - \bar{z}\,\Omega'(z) - \omega(z)\}$$

and hence the boundary conditions become

$$\begin{aligned} &t\,\Omega''(t) + t^2a^{-2}\,\omega'(t) - a^2t^{-1}\,\overline{\Omega''(t)} - a^2t^{-2}\,\overline{\omega'(t)} = 2\mathrm{i}\,g(t) \\ &\kappa a^2t^{-1}\,\Omega(t) - a^2\,\overline{\Omega'(t)} - a^2t^{-1}\,\overline{\omega(t)} \\ &\qquad + \kappa t\,\overline{\Omega(t)} - a^2\,\Omega'(t) - t\,\omega(t) = 4\mu a\,f(t) \end{aligned} \tag{4.96}$$

on $t\bar{t} = a^2$, where the functions $\Omega(z)$, $\omega(z)$ are defined for $z \in S^+$. Milne-Thomson,[41] Section 5.36 has derived the solution to this problem by using a method based on continuation, and this is illustrated in a more general context in Section 5.7. It is simpler to follow the direct method of Section 4.3 and to operate on each of the boundary conditions above with the integral

$$\frac{1}{2\pi\mathrm{i}}\int_\Gamma \frac{\mathrm{d}t}{t-z} \quad \text{for } z \in S^+.$$

For definiteness we shall suppose S^+ to be the region $|z| < a$, the solution for the region $|z| > a$ proceeding similarly. On noting in the first boundary condition that $z\,\Omega''(z)$, $z^2\omega'(z)$ are holomorphic in S^+ we see

$$\frac{1}{2\pi\mathrm{i}}\int_\Gamma \left\{t\,\Omega''(t) + \frac{t^2}{a^2}\,\omega'(t)\right\}\frac{\mathrm{d}t}{t-z} = z\,\Omega''(z) + \frac{z^2}{a^2}\,\omega'(z) \qquad (z \in S^+).$$

Similarly

$$\frac{1}{2\pi\mathrm{i}}\int_\Gamma \left\{\frac{1}{t}\,\overline{\Omega''(t)} + \frac{1}{t^2}\,\overline{\omega'(t)}\right\}\frac{\mathrm{d}t}{t-z} = \frac{1}{2\pi\mathrm{i}}\int_\Gamma \left\{\frac{1}{t}\,\overline{\Omega''\left(\frac{a^2}{\bar{t}}\right)} + \frac{1}{t^2}\,\overline{\omega'\left(\frac{a^2}{\bar{t}}\right)}\right\}\frac{\mathrm{d}t}{t-z}$$

and as the functions

$$\frac{1}{z}\,\overline{\Omega''\left(\frac{a^2}{\bar{z}}\right)} \qquad \text{and} \qquad \frac{1}{z^2}\,\overline{\omega'\left(\frac{a^2}{\bar{z}}\right)}$$

are holomorphic in S^- we may extend this integral to a contour integral over the circle $|z| = R$ for large R. Thus the integral may be shown to be zero. Hence the first boundary condition of (4.96) becomes

$$z\,\Omega''(z) + \frac{z^2}{a^2}\,\omega'(z) = \frac{1}{\pi}\int_\Gamma \frac{g(t)\,\mathrm{d}t}{t-z} \quad (z \in S^+). \tag{4.97}$$

Applying the same methods of evaluation to the second boundary condition of (4.96)

$$\begin{aligned} &\frac{1}{2\pi\mathrm{i}}\int_\Gamma \left\{\frac{\kappa a^2}{t}\,\Omega(t) - a^2\,\Omega'(t) - t\,\omega(t)\right\}\frac{\mathrm{d}t}{t-z} \\ &\qquad = \kappa a^2\left\{\frac{\Omega(z)}{z} - \frac{\Omega(0)}{z}\right\} - a^2\,\Omega'(z) - z\,\omega(z) \quad (z \in S^+) \end{aligned}$$

and

$$\frac{1}{2\pi i}\int_{\Gamma}\left(-a^2\,\overline{\Omega'(t)} - \frac{a^2}{t}\,\overline{\omega(t)} + \kappa t\,\overline{\Omega(t)}\right)\frac{dt}{t-z}$$
$$= \frac{1}{2\pi i}\int_{\Gamma}\left\{-\,a^2\,\overline{\Omega'}\left(\frac{a^2}{t}\right) - \frac{a^2}{t}\,\overline{\omega}\left(\frac{a^2}{t}\right) + \kappa t\,\overline{\Omega}\left(\frac{a^2}{t}\right)\right\}\frac{dt}{t-z}.$$

Again this latter integral may be expressed in terms of a contour integral around a circle of large radius R about the origin and hence shown to have the value

$$-a^2\,\overline{\Omega'(0)} + \kappa(z\,\overline{\Omega(0)} + a^2\,\overline{\Omega'(0)}) \quad \text{for } z \in S^+.$$

Thus the second boundary condition of (4.96) leads to the equation

$$\kappa a^2 z^{-1}\,\Omega(z) - a^2\,\Omega'(z) - z\,\omega(z) - \kappa a^2 z^{-1}\,\Omega(0) + \kappa z\,\overline{\Omega(0)}$$
$$+ (\kappa - 1)\,a^2\,\overline{\Omega'(0)} = \frac{4\mu a}{2\pi i}\int_{\Gamma}\frac{f(t)\,dt}{t-z}. \qquad (4.98)$$

The potentials $\Omega(z)$, $\omega(z)$ may now be found by solving (4.97) and (4.98) simultaneously. On eliminating $\omega'(z)$ we find $\Omega(z)$ satisfies the first-order differential equation

$$\Omega'(z) - \frac{2\kappa}{(\kappa+1)z}\,\Omega(z) = \frac{1}{\kappa+1}\left[\frac{1}{\pi}\int_{\Gamma}\frac{g(t)\,dt}{t-z} + \frac{2\mu z^2}{\pi i a}\frac{d}{dz}\left\{\int_{\Gamma}\frac{f(t)\,dt}{(t-z)z}\right\}\right.$$
$$\left. + (\kappa - 1)\,\overline{\Omega'(0)} - \frac{2\kappa}{z}\,\Omega(0)\right]$$
$$= F(z) \quad (z \in S^+). \qquad (4.99)$$

The integrating factor of the above equation is $z^{-\gamma}$, $\gamma = 2\kappa/(\kappa + 1)$ and hence we find

$$\Omega(z) = \alpha z^{\gamma} + z^{\gamma}\int z^{-\gamma}F(z)\,dz \qquad (4.100)$$

where α is an arbitrary constant. Since $1 < \kappa < 3$, the exponent $\gamma = 2\kappa/(\kappa + 1)$ takes a non-integral value in the range $1 < \gamma < 1{\cdot}5$ and the term αz^{γ} in (4.100) is multivalued. It will be seen that $F(z)$ in (4.99) may be represented in the form

$$-2\kappa\,\Omega(0)z^{-1} + \text{a power series in } z$$

and hence the second term in (4.100) is single valued. As $\Omega(z)$ must be single valued in S^+ we conclude $\alpha = 0$.

The remaining constants occurring in (4.100) are $\Omega'(0)$ and $\Omega(0)$ and we would expect to determine them by taking power series expansions about the origin in (4.100) or (4.99). In fact on taking these expansions $\Omega(0)$ remains unspecified, but we find

$$2(\kappa - 1)\mathrm{Re}\{\Omega'(0)\} = \frac{2\mu}{\pi \mathrm{i} a}\int_\Gamma \frac{f(t)\,\mathrm{d}t}{t} - \frac{1}{\pi}\int_\Gamma \frac{g(t)\,\mathrm{d}t}{t}. \tag{4.101}$$

This condition implies the right-hand side of (4.101) is real and hence, since $f(t)$ and $g(t)$ are real,

$$\int_\Gamma \frac{g(t)\,\mathrm{d}t}{t} = \mathrm{i}\int_0^{2\pi} \tau_{r\theta}\bigg|_{r=a} \mathrm{d}\theta = 0 \tag{4.102}$$

which is merely the condition that the resultant moment applied to the disc should be zero. Hence

$$(\kappa - 1)\mathrm{Re}\{\Omega'(0)\} = \frac{\mu}{\pi \mathrm{i} a}\int_\Gamma \frac{f(t)\,\mathrm{d}t}{t}. \tag{4.103}$$

In a similar manner $\omega(z)$ is defined from (4.98) and must be holomorphic at the origin. This condition implies

$$\Omega(0) = 0. \tag{4.104}$$

Thus only the imaginary part of $\Omega'(0)$ remains unspecified which implies $\Omega(z)$ is undefined to within a term of the form $\mathrm{i}Cz$ where C is real. It will be seen that $\omega(z)$ is completely defined by (4.98) and is independent of C. Thus the displacement field D is defined apart from a term of the form $(\kappa - 1)\mathrm{i}Cz/2\mu$ which corresponds to a rigid-body rotation of the disc about the origin, which is to be expected from the boundary conditions. The solution to the problem is now complete.

As a simple example we suppose

$$u_r = -\varepsilon, \qquad \tau_{r\theta} = 0. \tag{4.105}$$

Then as

$$\frac{1}{2\pi \mathrm{i}}\int_\Gamma \frac{-\varepsilon\,\mathrm{d}t}{t - z} = -\varepsilon \quad \text{for } z \in S^+$$

we find

$$F(z) = \frac{1}{\kappa + 1}\left\{(\kappa - 1)\,\overline{\Omega'(0)} + \frac{4\mu\varepsilon}{a}\right\}$$

$$(\kappa - 1)\mathrm{Re}\{\Omega'(0)\} = \frac{-2\mu\varepsilon}{a}.$$

Thus

$$\Omega(z) = \left(\frac{2\mu\varepsilon}{a} + \mathrm{i}C\right)\frac{z}{1 - \kappa}$$

and from (4.98) we find

$$\omega(z) = 0.$$

Hence, disregarding the rigid-body rotation, the solution is simply

$$\Omega(z) = \frac{2\mu\varepsilon z}{(1-\kappa)a}, \qquad \omega(z) = 0.$$

2. *Incomplete contact*

The cases where separation occurs between an insert or inclusion and the surrounding medium under either an applied loading or a non-uniform expansion of the inclusion have attracted much attention in view of their practical importance. For example if an infinite sheet under a uniaxial tension T along the y-axis contains a rigid circular inclusion then one might expect separation to occur along the arcs L', namely, $\phi \leqslant \theta \leqslant \pi - \phi$,

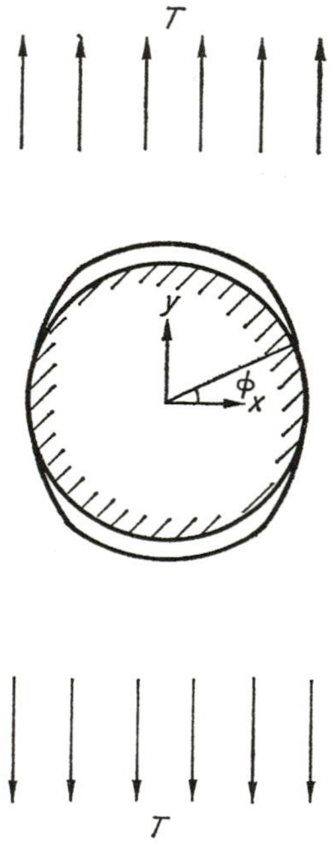

Figure 4.4

$\pi + \phi \leqslant \theta \leqslant 2\pi - \phi$ where ϕ is the (unknown) contact angle (see Figure 4.4) for sufficiently large values of T. Assuming the materials make a frictionless contact over the contact regions L: $|\theta| \leqslant \phi$, $|\theta - \pi| \leqslant \phi$

the boundary conditions for the infinite medium $|z| \geqslant a$ are

$$\tau_{rr} + \mathrm{i}\tau_{r\theta} = 0 \quad (t \in L')$$
$$u_r = \tau_{r\theta} = 0 \quad (t \in L).$$

We shall consider the slightly more general conditions

$$\begin{aligned} \tau_{rr} + \mathrm{i}\tau_{r\theta} &= 0 \qquad (t \in L') \\ \tau_{r\theta} &= 0 \qquad (t \in L) \\ u_r &= f(t) \quad (t \in L) \end{aligned} \tag{4.106}$$

where $f(t)$ is real and has the following symmetries

$$f(t) = f(-t) = f(\bar{t}). \tag{4.107}$$

These symmetries imply there is a zero resultant force over the hole $|z| = a$.

There are two equivalent methods of solution for this problem. The first is to use the condition $\tau_{r\theta} = 0$ on $|z| = a$ and to follow the direct method of the last section. This results in an equation similar to (4.97) relating $\omega'(z)$ to $a^2z^{-1}\,\Omega''(z)$ with some additional terms due to the fact that S^+ is now the region $|z| > a$. On eliminating $\omega'(z)$ the remaining boundary conditions $\tau_{rr} = 0$ $(t \in L')$ and $u_r = f(t)$ on L then yield a boundary value problem for $\Omega(z)$. It turns out that this boundary value problem is most easily reduced by continuing $\Omega(z)$ from S^+ into S^- by the stress continuation. We shall not investigate this method in detail as it is more direct to use the stress continuation from the beginning. Thus the alternative method of solution is to satisfy identically the stress-free conditions on L' by using the stress continuation defined in Section 4.4. In this case from (4.52) we find

$$\begin{aligned} 2\mu(u_r + \mathrm{i}u_\theta) &= \mathrm{e}^{-\mathrm{i}\theta}\{\kappa\,\Omega^+(t) + \Omega^-(t)\} \\ \tau_{rr} + \mathrm{i}\tau_{r\theta} &= \Omega'^+(t) - \Omega'^-(t) \end{aligned} \tag{4.108}$$

where $\Omega(z)$ is a holomorphic function in the plane cut along the contact regions L and behaves at the origin and infinity in accord with (4.49) and (4.50). On substituting these relations into the remaining boundary conditions of (4.106) we find

$$\left.\begin{aligned} \mathrm{Im}[\Omega'^+(t) - \Omega'^-(t)] &= 0 \\ \mathrm{Re}[\mathrm{e}^{-\mathrm{i}\theta}\{\kappa\,\Omega^+(t) + \Omega^-(t)\}] &= 2\mu f(t) \end{aligned}\right\} (t \in L) \tag{4.109}$$

where L is described in a clockwise direction. Unlike the corresponding problem of Section 3.8, as far as the author is aware it has not been possible to reduce these equations to a Hilbert problem. In view of this the following procedure has normally been adopted.† Assume $\tau_{rr} = N(t)$ $(t \in L)$

† For references see p. 124.

where the real function N satisfies the symmetries t of (4.107), then $\Omega'(z)$ satisfies the Cauchy problem

$$\Omega'^{+}(t) - \Omega'^{-}(t)\begin{cases} = 0 & (t \in L') \\ = N(t) & (t \in L) \end{cases}$$

and hence from Section 1.6 and (4.49), (4.50)

$$\Omega'(z) = \frac{1}{2\pi i}\int_L \frac{N(t)\,dt}{t - z} + \frac{T}{4}\left(1 + \frac{2a^2}{z^2}\right) \tag{4.110}$$

since $C + iD = T/2$, $A + iB = T/4$ and $X + iY = 0$ and we note L is described in a clockwise direction. It remains to satisfy the second condition of (4.109). As (4.110) defines $\Omega'(z)$ it is natural to differentiate (4.109) and express this condition in terms of $\Omega'(z)$. We find

$$\begin{aligned} &a^2(\kappa - 1)\{\phi'^{+}(t) - \phi'^{-}(t)\} - a^2(\kappa + 1)t\{\phi''^{+}(t) + \phi''^{-}(t)\} \\ &\quad = 4\mu a\left\{f(t) + \frac{\partial^2}{\partial\theta^2}f(t)\right\} + a^2(\kappa + 1)\left\{\alpha - \frac{T}{2} - \frac{3T}{2}\left(\frac{a^2}{t^2} + \frac{t^2}{a^2}\right)\right\} \end{aligned} \tag{4.111}$$

where

$$\phi'(z) = \frac{1}{2\pi i}\int_L \frac{N(t)\,dt}{t - z} \quad \text{and} \quad \alpha = \frac{1}{2\pi i}\int_L \frac{N(t)\,dt}{t}.$$

Hence using the Plemelj formulae (1.21) $N(t)$ satisfies the singular integral equation

$$\begin{aligned} (\kappa - 1)N(t) - (1 + \kappa)t\int_L \frac{N'(\tau)\,d\tau}{\tau - t} &= \frac{4\mu}{a}\{f(t) - tf'(t) - t^2 f''(t)\} \\ &\quad - (\kappa + 1)\left\{\alpha + \frac{T}{2} + \frac{3T}{2}\left(\frac{a^2}{t^2} + \frac{t^2}{a^2}\right)\right\}, \quad \text{for } t \in L. \end{aligned} \tag{4.112}$$

A similar equation has been derived by Sherementev[52] and Stippes, Wilson and Krull[62] in the case of separation of a circular elastic insert from the matrix in which it is embedded under the assumptions that the insert and matrix have different elastic constants but make a smooth contact. In fact (4.112) is a Prandtl type of singular integro-differential equation which also occurs in aerodynamic problems associated with wings of finite span. Equations of this type have been discussed by Muskhelishvili,[44] Chapter 17. A simple solution to (4.112) is obtainable only when $\kappa = 1$ which, in the case of plane strain, corresponds to the medium being incompressible, otherwise the solution must be obtained by numerical methods. Some calculations in the cases of circular and elliptic inserts have been performed by Wilson[77] and Goree,[18] and further references have been given by England.[11] It will be appreciated that this problem is of a different nature to those considered previously.

Examples on Chapter 4

1. Use the series method to show that the complex potentials in an infinite elastic medium $|z| \geqslant a$ with zero stresses at infinity due to a uniform normal displacement $u_r = \varepsilon$ and shear stress $\tau_{r\theta} = 0$ on the boundary $|z| = a$ are $\Omega(z) = 0, \quad \omega(z) = -2\mu a\varepsilon/z$.

2. Using the series method show that the complex potentials in an elastic disc $|z| \leqslant R$ due to the boundary conditions $u_r = \delta$, $\tau_{r\theta} = 0$ on $|z| = R$ are $\Omega(z) = \dfrac{2\mu\delta\, z}{(\kappa - 1)R}, \quad \omega(z) = 0.$

3. A smooth elastic disc of radius R is compressed and inserted into a hole of radius a in an infinite plate composed of a different elastic material. Assuming R is greater than a and $R - a$ is small show that the pressure across the interface of the media is

$$\frac{4\mu_1\mu_2}{\mu_2(\kappa_1 - 1) + 2\mu_1}\left(\frac{R}{a} - 1\right)$$

where μ_1, κ_1 are the elastic constants of the disc and μ_2, κ_2 of the infinite medium, and determine the radius of the interface.

4. Solve the displacement problem for the annulus $a \leqslant |z| \leqslant b$ when $D_2 = 0$ on $|z| = b$ and $D_1 = \varepsilon$ (real) on $|z| = a$. Confirm that the resultant force over the hole is

$$2\pi\varepsilon\mu\kappa(1 + \kappa)\left[\kappa^2 \log\left(\frac{b}{a}\right) - \left(\frac{b^2 - a^2}{b^2 + a^2}\right)\right]^{-1}.$$

5. If the stress distribution $\tau_{rr} + \mathrm{i}\tau_{r\theta} = -p/2a\alpha$ where p is real, is applied to the arcs $|\theta| < \alpha$, $|\theta - \pi| < \alpha$ of the surface $|z| = a$ of the disc $|z| \leqslant a$, the remainder of the surface being unstressed, use the direct method of solution of Section 4.3 to show that

$$\Omega'(z) = \frac{p}{2\pi a}\left[1 - \frac{1}{2\alpha\mathrm{i}}\log\left(\frac{a^2\mathrm{e}^{2\mathrm{i}\alpha} - z^2}{a^2\mathrm{e}^{-2\mathrm{i}\alpha} - z^2}\right)\right].$$

Hence deduce that the potential due to diametrically opposed point forces of magnitude p acting at $z = \pm a$ is

$$\Omega'(z) = \frac{p}{2\pi a}\frac{(z^2 + a^2)}{(z^2 - a^2)}.$$

6. Use the direct solution of Section 4.3 to satisfy the boundary conditions

$$\gamma\, \Omega(t) + t\,\overline{\Omega'(t)} + \overline{\omega(t)} = F(t) \quad \text{on } t = a\mathrm{e}^{\mathrm{i}\theta}$$

for the region $|z| \geqslant a$. If the stress distribution

$$\tau_{rr} + \mathrm{i}\tau_{r\theta} = -\frac{(X + \mathrm{i}Y)}{2\pi a}\mathrm{e}^{-\mathrm{i}\theta}$$

acts over the hole $|z| = a$ and there are zero stresses at infinity confirm that the complex potentials are given by (4.27).

7. Use the direct method of Section 4.3 to confirm that the potentials

$$\Omega'(z) = \frac{T}{4}\left(1 + \frac{2a^2}{\kappa z^2}\right), \qquad \omega'(z) = -\frac{T}{2}\left(1 + \frac{(\kappa - 1)a^2}{2z^2} - \frac{3a^4}{\kappa z^4}\right)$$

solve the problem of the deformation of an infinite medium $|z| \geqslant a$ containing a reinforced circular hole on which $D = 0$ under a uniform tension T at infinity acting parallel to the x-axis.

8. Use the equations (4.51) resulting from the stress continuation to solve example 5.

9. Use the equations resulting from (4.1) and the displacement continuation (4.53) to solve example 7.

10. Determine the complex potentials for a circular disc of radius a with an unstressed boundary in equilibrium under the point forces X and $-X$ at the internal points $z = 0$ and $z = \frac{1}{2}a$ respectively. Evaluate the hoop stress $\tau_{\theta\theta}$ on the boundary.

11. Determine the complex potentials for the circular disc $|z| \leqslant a$ corresponding to a point force X at $z = 0$ and an equal and opposite point force $-X$ applied at the boundary point $z = a$, the rest of the boundary remaining unstressed.

12. An infinite elastic medium $|z| \geqslant a$ is bonded over the whole of its boundary to a rigid inclusion of radius a. If the inclusion is acted on by a force $X + \mathrm{i}Y$ and moment M about its centre show that the inclusion will rotate through a small angle $\varepsilon = M/4\pi\mu a^2$ and that

$$\Omega'(z) = -\frac{(X + \mathrm{i}Y)}{2\pi(1 + \kappa)}\frac{1}{z}, \qquad \omega'(z) = \frac{\kappa(X - \mathrm{i}Y)}{2\pi(1 + \kappa)}\frac{1}{z} - \frac{\mathrm{i}M}{2\pi z^2} - \frac{(X + \mathrm{i}Y)}{\pi(1 + \kappa)}\frac{a^2}{z^3}$$

13. Use the method of Section 4.5 for the annulus to solve the displacement problem defined in example 4. Hence evaluate the stress on $|z| = b$.

14. In the illustration of Section 4.6 evaluate the stresses $\tau_{rr} + i\tau_{r\theta}$ over the reinforced arc L and investigate the form of the stress singularities at the ends $z = ae^{\pm i\phi}$. Examine also the values of $\partial D/\partial\theta$ on $|z| = a$.

15. An infinite medium $|z| \geqslant a$, with zero stress and rotation at infinity contains a circular hole. An arc $|\theta| \leqslant \phi$ of the hole is strongly reinforced so that it moves as a rigid body and is acted on by a force which is symmetrically distributed about $\theta = 0$, and has the resultant X. If the rest of the hole is unstressed, use the method of Section 4.6 to show that

$$\Omega'(z) = -\frac{X}{2\pi(1+\kappa)}\left(\frac{a\kappa e^{-2\phi\beta}}{z} + 1\right)(z - ae^{i\phi})^{-\frac{1}{2}-i\beta}(z - ae^{-i\phi})^{-\frac{1}{2}+i\beta}.$$

16. Use the theory of Section 4.7 to calculate the complex potentials corresponding to the inflation of a semicircular arc crack by a uniform pressure $\tau_{rr} = -p$ on the edges of the crack, the infinite medium having zero stress and rotation at infinity. How does this solution relate to the case of a plane under an all-round tension at infinity, cut along a semicircular arc?

17. The problem of a circular elastic inclusion bonded to a different elastic material except over a set of arc cracks may also be solved using the methods of Section 4.7. Assume the elastic medium $|z| \geqslant a$ has coefficients μ_1, κ_1, complex potentials $\Omega_1(z)$, $\omega_1(z)$ and is bonded (displacements and normal stresses continuous) to the inclusion $|z| \leqslant a$, with coefficients μ_2, κ_2 and potentials $\Omega_2(z)$, $\omega_2(z)$ except over the arcs L of $|z| = a$.

Show that the continuity conditions across $|z| = a$ are satisfied if we put

$$(\mu_1 + \mu_2\kappa_1)\Omega_1(z) = \mu_1\,\theta(z) + \phi(z),$$
$$\omega_1(z) = \overline{\Omega_2\left(\frac{a^2}{\bar{z}}\right)} - \frac{a^2}{z}\,\Omega_1'(z) - \overline{\theta\left(\frac{a^2}{\bar{z}}\right)}$$
$$(\mu_2 + \mu_1\kappa_2)\Omega_2(z) = \mu_2\,\theta(z) + \phi(z),$$
$$\omega_2(z) = \overline{\Omega_1\left(\frac{a^2}{\bar{z}}\right)} - \frac{a^2}{z}\,\Omega_2'(z) - \overline{\theta\left(\frac{a^2}{\bar{z}}\right)}$$

where $\theta(z)$, $\phi(z)$ are holomorphic functions in the plane cut along L with the possible exclusion of the origin and infinity. If equal and opposite normal pressures $\tau_{rr} = -p$ are applied to the edge of the arc cracks L show that $\theta(z)$, $\phi(z)$ satisfy boundary conditions

$$\theta'^{+}(t) - \theta'^{-}(t) = 0$$
$$\phi'^{+}(t) + \alpha\phi'^{-}(t) = -\tfrac{1}{2}(\mu_1 + \mu_2\kappa_1)p - \mu_1(1 - \alpha\kappa_2)\theta'(t)$$

on L where $\alpha = (\mu_1 + \mu_2\kappa_1)/(\mu_2 + \mu_1\kappa_2)$.

Confirm that the solution to this problem for a single crack on $|\theta| \leqslant \phi$

with zero stresses at infinity has the form

$$\theta(z) = Az$$
$$\phi(z) = Bz + C(z - ae^{i\phi})^{\frac{1}{2}+i\beta}(z - ae^{-i\phi})^{\frac{1}{2}-i\beta}$$

where A, B, C are constants and $2\pi\beta = \log \alpha$. Hence show that the distance apart of corresponding points on the edges of the crack is given by

$$\frac{a(1 + \alpha)Ce^{-\beta\phi}}{2\mu_1\mu_2\alpha^{1/2}} |e^{i\theta} - e^{i\phi}|^{\frac{1}{2}+i\beta}|e^{i\theta} - e^{-i\phi}|^{\frac{1}{2}-i\beta} e^{\frac{1}{2}i\theta}$$

and confirm that this oscillates taking negative values as $\theta \to \pm\phi$.

18. Use both the series method and the direct method of solution to find the stress distribution in a plate containing a circular hole where the displacement at the point $t = ae^{i\theta}$ on the boundary is given by

$$u_r + iu_\theta = \varepsilon \sin n\theta$$

where n is an integer.

19. For an infinite medium $|z| \geqslant a$ containing an unstressed circular hole, under the action of the principal stresses N_1, N_2 at infinity where N_1 makes an angle α with the x-axis, evaluate the hoop stress $\tau_{\theta\theta}$ on the boundary of the hole. Find where the maxima and minima of $\tau_{\theta\theta}$ occur.

20. A circular disc contains a set of point forces and moments which if acting in an infinite medium would correspond to the complex potentials $\Omega_0(z)$, $\omega_0(z)$. If the boundary of the disc is held so that $D = 0$ at all points on the boundary, determine the complex potentials in the disc. In particular examine the case of a single point force acting at the origin and evaluate the stresses on the boundary.

21. A rigid insert of radius $a + \varepsilon$ is fitted into a central hole of radius a in an elastic disc of radius b. If the insert and the elastic disc make a frictionless contact, find the stress distribution in the disc when a force X is applied to the rigid insert, assuming the outer rim of the disc is held so that $D = 0$. Will separation ever occur between the insert and the disc?

5

REGIONS WITH CURVILINEAR BOUNDARIES

Introduction

In the theory of two-dimensional linear elasticity one of the most useful techniques for the solution of boundary value problems for awkwardly shaped regions is to transform the region into one of simpler shape. In general, more cumbersome boundary conditions are produced by the transformation but the difficulties of handling these are outweighed by the simpler geometry that results.

In this chapter we consider the basic boundary value problems for regions which map conformally onto the interior, or exterior, of a circle. Such regions are either discs bounded by a contour C or an infinite plane containing a hole having C as its boundary, where the curvature of C is continuous. As the boundary conditions for these problems transform into similar conditions on the circle, the three basic methods of solution discussed in Chapter 4 may be applied. Of these we shall only consider in detail the direct method and the method based on continuation.

5.1 Conformal transformations

In this section we summarize the main properties of conformal transformations which will be used in the text. The usefulness of conformal transformations for these problems derives not from the fact that they are a wide class of transformations, but that they enable us to extend the basic complex variable formulation to the transformed problem, and hence to apply the powerful methods of solution derived in the earlier sections for the half-plane and circular regions.

We consider the transformation $z = m(\zeta)$ which maps the points of a region R_ζ in the ζ-plane into points of a region R_z in the z-plane. If we select a single-valued branch of the function $z = m(\zeta)$ then specification of the region R_ζ defines the region R_z. We shall assume the transformation between R_ζ and R_z is one to one and invertible so that by selecting a particular branch of the inverse transformation $\zeta = m^{-1}(z)$ we may define the corresponding one-to-one mapping from R_z onto R_ζ. Now to preserve the basic complex variable formulation we assume in addition that $m(\zeta)$ is holomorphic in R_ζ (it has already been assumed single valued) and that $m'(\zeta) \neq 0$ for all $\zeta \in R_\zeta$.

The properties of this type of transformation may be most easily investigated by examining the relationship between corresponding curves in R_ζ and R_z. Suppose L_ζ is an arc in R_ζ which passes through the point $\zeta_0 \in R_\zeta$, then as ζ describes L_ζ the corresponding point $z = m(\zeta)$ will describe an arc L_z passing through the point $z_0 = m(\zeta_0) \in R_z$. Since $m(\zeta)$ is holomorphic the relation between a small increment $\delta\zeta$ along L_ζ at ζ_0 and the corresponding increment δz along L_z is

$$\delta z = m'(\zeta_0)\, \delta\zeta.$$

From this relation various properties of the mapping may be deduced. First, provided $m'(\zeta_0) \neq 0$, the mapping is locally one to one and invertible giving

$$\delta\zeta = \frac{\delta z}{m'(\zeta_0)}.$$

Secondly, since

$$\left|\frac{dz}{d\zeta}\right| = |m'(\zeta_0)|$$

the local magnification is $|m'(\zeta_0)|$ and is independent of the direction of L_ζ at $\zeta = \zeta_0$. Also

$$\arg(\delta z) = \phi + \arg(\delta\zeta) \tag{5.1}$$

where $\phi = \arg(m'(\zeta_0))$ so that the angle of inclination of the element δz of L_z to the real axis in the z-plane is equal to the angle of inclination of the element $\delta\zeta$ to the real axis of the ζ-plane plus a constant angle. Thus the neighbourhood of the point ζ_0 in the ζ-plane is transformed by means of a magnification $|m'(\zeta_0)|$ and rigid-body rotation through the angle $\phi = \arg(m'(\zeta_0))$ into the neighbourhood of the point z_0 in the z-plane provided $m'(\zeta_0) \neq 0$. In particular if two arcs L_ζ^1, L_ζ^2 intersect at $\zeta = \zeta_0$ with an angle of intersection τ, then the corresponding arcs L_z^1, L_z^2 in R_z meet at z_0 in the same angle τ and with the same orientation. A transformation which preserves angles in this manner is said to be *conformal*, and the transformation from R_ζ to R_z is conformal when $m(\zeta)$ is holomorphic and $m'(\zeta) \neq 0$ for all ζ in R_ζ. These conditions also ensure the inverse transformation from R_z to R_ζ is conformal.

We shall be concerned in this chapter only with regions R_z which map conformally onto the interior or exterior of the circle $|\zeta| = a$ in the ζ-plane.

As the circles $|\zeta| =$ constant and the radial lines $\arg \zeta =$ constant, form an orthogonal curvilinear net in the ζ-plane, the conformal map $z = m(\zeta)$ of these curves forms an orthogonal curvilinear net in the region R_z of the z-plane. In fact the maps of the curves $|\zeta| = \xi$, $\arg \zeta = \eta$ may be used to define a system of orthogonal curvilinear coordinates (ξ, η) in the region R_z. In particular we see the curve $\xi = a$ is the boundary contour C, and, if R_ζ is the region $|\zeta| \leqslant a$, each curve $\xi = \xi_0$ lies entirely within the region R_z when ξ_0 is constant and $0 \leqslant \xi_0 \leqslant a$. Also the curves $\eta =$ constant, cut the boundary contour C, and every other curve of the family $\xi =$ constant, orthogonally in R_z. This property enables us to calculate the angle of inclination α of the normal to the coordinate curve $\xi =$ constant, at each point in terms of the mapping (see Figures 2.6 and 5.1). At the point z of R_z which corresponds to $\zeta = \xi_0 e^{i\eta}$ of R_ζ, the normal to the curve $\xi = \xi_0$ coincides with the map of the radial line $\arg \zeta = \eta$ from the ζ-plane. Hence from (5.1)

$$\alpha = \arg(m'(\zeta)) + \eta, \qquad \zeta = \xi_0 e^{i\eta}$$

so that

$$e^{2i\alpha} = e^{2i\eta} \frac{m'(\zeta)}{\overline{m'(\zeta)}} = \frac{\zeta\, m'(\zeta)}{\overline{\zeta}\, \overline{m'(\zeta)}}. \tag{5.2}$$

It will be seen that the conformal mapping defines a natural coordinate system in the region R_z.

The question which now arises is how general may a region R_z be? It has been proved[6] that it is possible to conformally map a simply connected region R_z bounded by a simple contour C with a continuously turning tangent onto the region $|\zeta| \leqslant a$ by a mapping $z = m(\zeta)$ which is such that $m'(\zeta) \neq 0$ at the interior points $|\zeta| < a$. In the following sections it is important to know whether $m'(\zeta) = 0$ at some point on the boundary $|\zeta| = a$. It may be shown that $m'(\zeta) \neq 0$ on the boundary $|\zeta| = a$ provided the boundary contour C of R_z has a continuously changing curvature. In the sequel we shall assume C satisfies such a condition. Although the discussion above refers to the region $|\zeta| \leqslant a$ it is possible (and more natural) to map the infinite plane containing a single hole bounded by the contour C onto the region $|\zeta| \geqslant a$ in such a way that $m'(\zeta) \neq 0$ for $|\zeta| \geqslant a$ if C has a continuous curvature. This may be seen by introducing the subsidiary mapping $\zeta = a^2/\zeta_1$ taking the region $|\zeta_1| \leqslant a$ conformally onto the region $|\zeta| \geqslant a$.

In general the transformation $z = m(\zeta)$ between R_z and R_ζ is only determined to within a rigid-body displacement, so that if we specify the correspondence of two arbitrarily chosen points $\zeta_0 \in R_\zeta$ and $z_0 \in R_z$ and arbitrarily chosen directions at ζ_0 and z_0, the transformation is determined uniquely. Normally it proves convenient to map bounded simply connected regions onto $|\zeta| \leqslant a$, and infinite regions onto $|\zeta| \geqslant a$. The particular forms of these transformations may be determined from Laurent's theorem since $m(\zeta)$ is holomorphic and single valued in either $|\zeta| \leqslant a$ or $|\zeta| \geqslant a$.

If R_z is a bounded simply connected region occupying the interior of the contour C then it may be conformally mapped onto the region $|\zeta| \leqslant a$ by $z = m(\zeta)$ where $m(\zeta)$ has an expansion about $\zeta = 0$ of the form

$$z = m(\zeta) = a_0 + a_1\zeta + \cdots + a_n\zeta^n + \cdots \quad (a_1 \neq 0). \tag{5.3}$$

We can arrange that $a_0 = 0$ and that a_1 is real but non-zero by a choice of the ζ-axes (if $a_1 = 0$ then $m'(0) = 0$ and the map is not conformal).

Similarly if R_z is the region exterior to C it may be conformally mapped onto $|\zeta| \geqslant a$ by $z = m(\zeta)$ where from Laurent's theorem

$$z = m(\zeta) = a_1\zeta + a_0 + \frac{b_1}{\zeta} + \frac{b_2}{\zeta^2} + \cdots + \frac{b_n}{\zeta^n} + \cdots.$$

Terms in ζ^n for $n \geqslant 2$ are excluded as we have assumed $m(\zeta)$ is holomorphic at infinity. We can now, by choice of the ζ-axes, arrange that $a_0 = 0$ and a_1 is real so that

$$z = R\zeta + \frac{b_1}{\zeta} + \frac{b_2}{\zeta^2} + \cdots + \frac{b_n}{\zeta^n} + \cdots. \tag{5.4}$$

The truncation of the series (5.3) or (5.4) to n terms affords an approximation to the mapping function $z = m(\zeta)$ of known accuracy and defines a region R_z^n of approximately the same shape as R_z. It may be shown[43] under fairly general conditions that the solutions to elastic boundary value problems for the region R_z^n tend to the solutions for the region R_z as $n \to \infty$. This property has been used by Savin[50] and other authors to derive approximate solutions to several important problems.

So far we have restricted consideration to simply connected regions. In general it is only possible to conformally map R_z onto another region of like connectivity, for example, a ring-shaped region lying between two non-intersecting contours C_1, C_2 may be conformally mapped onto an annulus $a_1 \leqslant |\zeta| \leqslant a_2$. In general, problems involving multiply connected regions are very difficult to solve and will be omitted from this text. References to some particular solutions are given in Section 5.8.

We shall now briefly examine some mappings which are used for illustration in the text. In general it simplifies the algebra to perform mappings onto the interior or exterior of the unit circle $|\zeta| = 1$ and this procedure will be adopted.

1. The region $|\zeta| \geqslant 1$ may be mapped conformally onto the exterior of an ellipse by

$$z = m(\zeta) = R\left(\zeta + \frac{m}{\zeta}\right) \tag{5.5}$$

where $m'(\zeta) \neq 0$ for $|\zeta| \geqslant 1$ provided $|m| < 1$. Since the ellipse corresponds to the unit circle $\zeta = e^{i\eta}$ we see it has the parametric equation

$$z = R(1 + m)\cos\eta + iR(1 - m)\sin\eta$$

and thus has semi-axes $R(1 + m)$, $R(1 - m)$. Clearly to represent all ellipses it is sufficient to take m to lie in the range

$$0 \leqslant m < 1.$$

When $m = 0$ the ellipse is a circle of radius R, and in the limiting case when $m = 1$ it degenerates to a straight slit or crack of length $4R$. An examination of the curvilinear coordinate system defined by the mapping will show it is composed of orthogonal families of ellipses and hyperbolae. Clearly the region between two ellipses of this family may be transformed by (5.5) onto an annulus (see Figure 5.1).

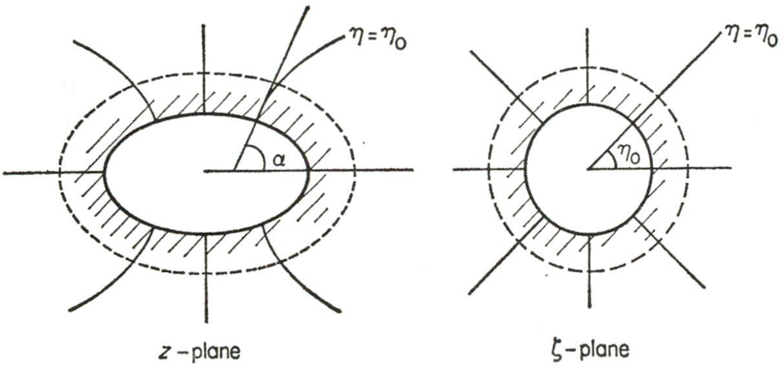

Figure 5.1

2. The region $|\zeta| \geqslant 1$ is mapped conformally onto the exterior region of a hypotrochoid by

$$z = m(\zeta) = R\left(\zeta + \frac{m}{\zeta^n}\right) \quad \left(0 \leqslant m < \frac{1}{n}\right) \tag{5.6}$$

where n is a positive integer and $m'(\zeta) \neq 0$ for $|\zeta| \geqslant 1$. The parametric equation of the hypotrochoid follows on putting $\zeta = e^{i\eta}$ and is

$$z = R(e^{i\eta} + me^{-in\eta}).$$

A hypotrochoid is the locus of a point fixed relative to a moving circle which rolls without slipping upon the interior of a fixed circle. The restriction $0 \leqslant m < 1/n$ above is equivalent to the assumption that the point lies in the interior of the moving circle, so that the hypotrochoid does not have loops and only has cusps if $m = 1/n$.

The main reason for interest in this mapping is that the general shapes of the hypotrochoids are curvilinear polygons; for $n = 2$ a curvilinear triangle, for $n = 3$ a curvilinear square etc. (see Figure 5.2), and hence

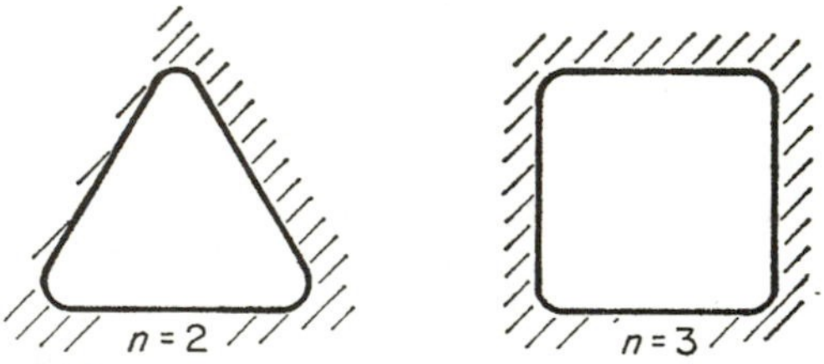

Figure 5.2

approximate regions of physical interest. The nature of the approximation will be examined shortly.

3. It might be thought that the conformal mapping of the interior of an ellipse onto the interior of a circle would be simple but this is not the case (see Kober,[30] p. 177, and Muskhelishvili,[43] Section 64). Instead we consider the representative transformation

$$z = R(\zeta + m\zeta^n) \quad \left(0 \leqslant m < \frac{1}{n}\right) \tag{5.7}$$

which maps the interior $|\zeta| \leqslant 1$ onto the interior of the epitrochoid

$$z = R(e^{i\eta} + me^{in\eta})$$

An epitrochoid is the locus of a point fixed relative to a moving circle which rolls without slipping on the exterior of a fixed circle. As for the hypotrochoid, the restriction $0 \leqslant m < 1/n$ ensures the point lies in the interior of the moving circle and the epitrochoid has no loops. For $m = 1/n$ the point lies on the surface of the moving circle and the epitrochoid has $n - 1$ cusps.

4. So far consideration has been directed towards mappings which are conformal up to and including the boundary $|\zeta| = 1$ of the region R_ζ. In particular this implies the curvature of the boundary contour C of R_z is continuous. It proves possible to relax these conditions and map the interior or exterior of a closed polygon onto the interior or exterior of the circle $|\zeta| = 1$. This mapping is effected by the Schwarz-Christoffel transformation,[45] and is conformal except at certain boundary points. Let us consider a simple closed polygon of n sides with exterior angles $\alpha_1, \alpha_2, \ldots, \alpha_n$ so that

$$\alpha_1 + \alpha_2 + \cdots + \alpha_n = 2\pi$$

(see Figure 5.3). If the vertices of the polygon are at the points $z_1, z_2, \ldots,$

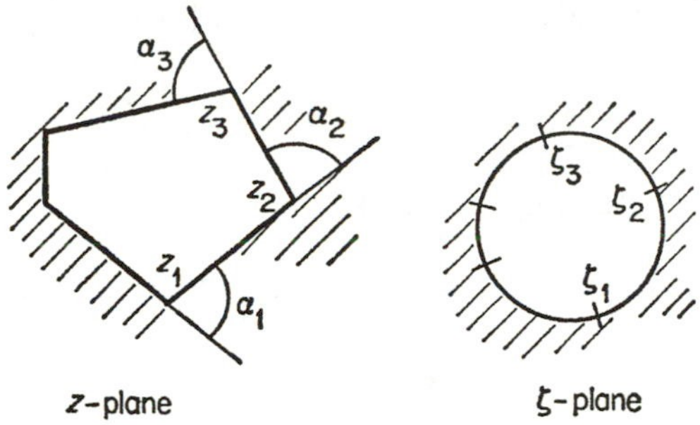

Figure 5.3

z_n, these points may be made to correspond with the set of points $\zeta_k = e^{i\eta_k}$ $(k = 1, 2, \ldots, n)$ on the circle $|\zeta| = 1$. The region outside the polygon may be mapped onto the region $|\zeta| \geqslant 1$ by the transformation $z = m(\zeta)$ where

$$m'(\zeta) = A \prod_{k=1}^{n} \left(1 - \frac{\zeta_k}{\zeta}\right)^{\alpha_k/\pi} \tag{5.8}$$

and A is a constant. The proof of this property has been given in several texts. It will be noticed that $m'(\zeta)$ may be expanded in a polynomial in $1/\zeta$ of the form

$$m'(\zeta) = A\left(1 - \frac{1}{\pi\zeta} \sum_{1}^{n} \alpha_k \zeta_k + O\left(\frac{1}{\zeta^2}\right)\right)$$

and hence $z = m(\zeta)$ will contain a term in $\log \zeta$ and be many valued unless the otherwise arbitrary points ζ_k are chosen so that $\sum_{k=1}^{n} \alpha_k \zeta_k = 0$.

The mapping defined by (5.8) is conformal at all interior points, but is not conformal at the boundary points ζ_k where $m'(\zeta_k) = 0$. It will be seen later that such points give rise to difficulties in the solution, and in many cases it has been found convenient to approximate (5.8) by a polynomial in $1/\zeta$ of finite degree. This is equivalent to replacing the polygonal hole by one of approximately the same shape, but with a smooth boundary, having small radii of curvature near the vertices.

It is of interest to examine the connection between this mapping and that for the hypotrochoidal hole (5.6). For simplicity we consider a regular polygon of n sides. In this case $\alpha_k = 2\pi/n$ for $k = 1, 2, \ldots, n$ and to preserve symmetry we take the points ζ_k to be the roots of the equation $\zeta^n = 1$. Then from (5.8)

$$m'(\zeta) = A\prod_{k=1}^{n}\left(1 - \frac{\zeta_k}{\zeta}\right)^{2/n} = \frac{A}{\zeta^2}\prod_{k=1}^{n}(\zeta - \zeta_k)^{2/n}$$

$$= \frac{A}{\zeta^2}(\zeta^n - 1)^{2/n} = A(1 - \zeta^{-n})^{2/n}.$$

On expanding as a power series in $1/\zeta$ for large $|\zeta|$

$$m'(\zeta) = A\left(1 - \frac{2}{n\zeta^n} + \frac{2(2 - n)}{2n^2}\frac{1}{\zeta^{2n}} + \cdots\right)$$

so that the mapping is approximated by

$$z = m(\zeta) = A\left(\zeta + \frac{2}{n(n-1)}\cdot\frac{1}{\zeta^{n-1}} - \frac{2(2-n)}{2n^2(2n-1)}\cdot\frac{1}{\zeta^{2n-1}} + \cdots\right) \tag{5.9}$$

to within a constant. Thus for an equilateral triangular hole ($n = 3$) the mapping is

$$z = m(\zeta) = A\left(\zeta + \frac{1}{3\zeta^2} + \frac{1}{45\zeta^5} + \cdots\right)$$

for a square hole ($n = 4$)

$$z = m(\zeta) = A\left(\zeta + \frac{1}{6\zeta^3} + \frac{1}{56\zeta^7} + \cdots\right)$$

the constant A merely affecting the scale and orientation of the polygon. The mapping may be approximated to any degree of accuracy by the truncation of the series (5.9). On comparing (5.6) and (5.9) we see that the mapping for a hypotrochoidal hole corresponds to the two-term approximation for a regular polygon of s sides when $n = s - 1$ and $m = 2/s(s - 1)$ in (5.6).

The properties of conformal transformations summarized above are considered in detail in most texts on functions of a complex variable. In particular the reader is referred to Churchill[8] and Carrier, Krook and Pearson.[7] For a compendium of conformal transformations see Kober.[30] Approximate methods of constructing such mappings have been considered by Kantorowich and Krylov.[26]

5.2 Basic formulae

In this section the formulae for the stresses and displacements in the body occupying the region R_z and the boundary conditions on the contour C are referred to the corresponding points in the image region R_ζ.†

As the conformal transformation $z = m(\zeta)$ defines a system of orthogonal curvilinear coordinates (ξ, η) in the region R_z occupied by the body by means of the relation $z = m(\zeta)$, $\zeta = \xi e^{i\eta}$ it is convenient to refer the stresses and displacements to these coordinates (see Figure 5.1). In particular following the definitions of Section 2.12 the normal and shear stresses across the curves ξ = constant, and in particular on the boundary C are denoted by $\tau_{\xi\xi}$, $\tau_{\xi\eta}$ respectively and the tangential or hoop stress by $\tau_{\eta\eta}$. Similarly the components of the displacement vector normal and tangential to the curves ξ = constant are u_ξ and u_η taken in the directions of increasing ξ and η. On using the results of Section 2.12 the stresses and displacements are related to the complex potentials by

$$\begin{aligned} \tau_{\xi\xi} + \tau_{\eta\eta} &= 2\{\Omega'(z) + \overline{\Omega'(z)}\} \\ \tau_{\xi\xi} + i\tau_{\xi\eta} &= \Omega'(z) + \overline{\Omega'(z)} - \{z\,\overline{\Omega''(z)} + \overline{\omega'(z)}\}e^{-2i\alpha} \end{aligned} \tag{5.10}$$

$$2\mu(u_\xi + iu_\eta) = 2\mu De^{-i\alpha} = (\kappa\,\Omega(z) - z\,\overline{\Omega'(z)} - \overline{\omega(z)})e^{-i\alpha} \tag{5.11}$$

where α has been defined in the previous section by (5.2). Similarly the resultant force $R(z)$ over an arbitrary arc in the body is given by the relation (see Section 2.7)

$$\Omega(z) + z\,\overline{\Omega'(z)} + \overline{\omega(z)} = i\,R(z) + \text{constant}. \tag{5.12}$$

These relations may be expressed in terms of the ζ coordinates by writing

$$\Omega(z) = \Omega(m(\zeta)) \equiv \Omega(\zeta), \qquad \omega(z) = \omega(m(\zeta)) \equiv \omega(\zeta)$$

and noting

$$\Omega'(z) = \Omega'(\zeta)\frac{d\zeta}{dz} = \frac{\Omega'(\zeta)}{m'(\zeta)}, \qquad \omega'(z) = \frac{\omega'(\zeta)}{m'(\zeta)}. \tag{5.13}$$

† Alternatively this process may be thought of as expressing the formulae in terms of the natural curvilinear coordinate system defined in R_z by the mapping.

Then on using (5.2)

$$\tau_{\xi\xi} + \tau_{\eta\eta} = 2\left\{\frac{\Omega'(\zeta)}{m'(\zeta)} + \overline{\frac{\Omega'(\zeta)}{m'(\zeta)}}\right\}$$

$$\tau_{\xi\xi} + \mathrm{i}\tau_{\xi\eta} = \frac{\Omega'(\zeta)}{m'(\zeta)} + \overline{\frac{\Omega'(\zeta)}{m'(\zeta)}} - \left[\frac{m(\zeta)}{m'(\zeta)}\frac{\mathrm{d}}{\mathrm{d}\zeta}\overline{\left\{\frac{\Omega'(\zeta)}{m'(\zeta)}\right\}} + \overline{\frac{\omega'(\zeta)}{m'(\zeta)}}\right]\frac{\zeta\, m'(\zeta)}{\overline{\zeta\, m'(\zeta)}} \tag{5.14}$$

and similarly

$$2\mu D = \kappa\, \Omega(\zeta) - \frac{m(\zeta)}{\overline{m'(\zeta)}}\overline{\Omega'(\zeta)} - \overline{\omega(\zeta)} \tag{5.15}$$

so that

$$2\mu|\zeta|\,|m'(\zeta)|(u_\xi + \mathrm{i}u_\eta) = \overline{\zeta\, m'(\zeta)}\left\{\kappa\,\Omega(\zeta) - \frac{m(\zeta)}{\overline{m'(\zeta)}}\overline{\Omega'(\zeta)} - \overline{\omega(\zeta)}\right\}. \tag{5.16}$$

Finally from (5.12) the resultant force formula becomes

$$\Omega(\zeta) + \frac{m(\zeta)}{\overline{m'(\zeta)}}\overline{\Omega'(\zeta)} + \overline{\omega(\zeta)} = \mathrm{i}\, R(m(\zeta)) + \text{constant}. \tag{5.17}$$

For a bounded simply connected region mapped onto the interior of the circle $|\zeta| = 1$ the new potentials $\Omega(\zeta) \equiv \Omega(m(\zeta))$ and $\omega(\zeta) \equiv \omega(m(\zeta))$ are holomorphic and single valued in $|\zeta| \leqslant 1$. In the case of an infinite region which is mapped conformally onto the region $|\zeta| \geqslant 1$ by the transformation (5.4)

$$z = R\zeta + \frac{b_1}{\zeta} + \frac{b_2}{\zeta^2} + \cdots$$

the potentials $\Omega(\zeta)$, $\omega(\zeta)$ are holomorphic in $|\zeta| \geqslant 1$ and behave as

$$\begin{aligned}\Omega(\zeta) &= (A + \mathrm{i}B)R\zeta - \frac{(X + \mathrm{i}Y)}{2\pi(1 + \kappa)}\log\zeta + \Omega_0(\zeta)\\ \omega(\zeta) &= (C + \mathrm{i}D)R\zeta + \frac{\kappa(X - \mathrm{i}Y)}{2\pi(1 + \kappa)}\log\zeta + \omega_0(\zeta)\end{aligned} \tag{5.18}$$

from (2.57) and (2.60) where $\Omega_0(\zeta)$, $\omega_0(\zeta)$ are holomorphic and single valued in $|\zeta| \geqslant 1$ and are bounded as $|\zeta| \to \infty$. In (5.18) $X + \mathrm{i}Y$ is the resultant force over the hole and $A + \mathrm{i}B$, $C + \mathrm{i}D$ are related to the stresses and rotation at infinity in the body occupying the region R_z by (2.61), (2.62).

The various types of boundary conditions which may hold on the boundary C of R_z may also be expressed as boundary conditions on the circle $|\zeta| = 1$. If the displacement $u + \mathrm{i}v = D(z)$ is specified on C then the corresponding condition on $|\zeta| = 1$ is simply, from (5.15),

$$\kappa\, \Omega(\zeta) - \frac{m(\zeta)}{m'(\zeta)}\, \overline{\Omega'(\zeta)} - \overline{\omega(\zeta)} = 2\mu\, D(m(\zeta)) \equiv 2\mu\, D(\zeta) \qquad (5.19)$$

where $D(\zeta)$ is a known function on the boundary $|\zeta| = 1$. Similarly the stress boundary value problem has the boundary conditions (5.12) on C where $R(z)$ is the resultant force over an arbitrary arc of C and is a known function of position. Thus from (5.17) it reduces to the following boundary condition on $|\zeta| = 1$

$$\Omega(\zeta) + \frac{m(\zeta)}{m'(\zeta)}\, \overline{\Omega'(\zeta)} + \overline{\omega(\zeta)} = \mathrm{i}\, R(m(\zeta)) + \text{constant} \qquad (5.20)$$

where $R(m(\zeta)) = R(\zeta)$ is a known function on $|\zeta| = 1$. For example if the normal stresses $\tau_{\xi\xi} + \mathrm{i}\tau_{\xi\eta}$ are specified on C then

$$R(z) = \int_A^z (\tau_{\xi\xi} + \mathrm{i}\tau_{\xi\eta})\mathrm{e}^{\mathrm{i}\alpha}\, \mathrm{d}s = -\mathrm{i} \int_A^z (\tau_{\xi\xi} + \mathrm{i}\tau_{\xi\eta})\, \mathrm{d}z$$

the integral being taken along C from a fixed point A in the direction so that the elastic medium R_z lies to the left. In the particular case of uniform pressure $\tau_{\xi\xi} + \mathrm{i}\tau_{\xi\eta} = -p$ then $R(z) = ipz$ so that

$$R(\zeta) = ip\, m(\zeta).$$

Clearly as in Section 4.1 the stress and displacement boundary value problems may be examined simultaneously by writing the boundary conditions (5.19) and (5.20) in the form

$$\gamma\, \Omega(t) + \frac{m(t)}{m'(t)}\, \overline{\Omega'(t)} + \overline{\omega(t)} = F(t), \quad \text{on } t = \mathrm{e}^{\mathrm{i}\eta} \qquad (5.21)$$

where $\gamma = 1$ for the stress problem, $\gamma = -\kappa$ for the displacement problem. For a bounded simply connected region, $F(t)$ has the forms given by (5.19) and (5.20) and the functions $\Omega(\zeta)$, $\omega(\zeta)$ are single valued in $|\zeta| \leqslant 1$. For an infinite region it is convenient to remove the logarithmic terms from the potentials by (5.18) with the resulting boundary conditions

$$\gamma\, \Omega_0(t) + \frac{m(t)}{m'(t)}\, \overline{\Omega_0'(t)} + \overline{\omega_0(t)} = F_0(t) \qquad (5.22)$$

where for the displacement problem

$$F_0(t) = -2\mu\, D(t) - \frac{\kappa(X + \mathrm{i}Y)}{2\pi(1 + \kappa)} \log(t\bar{t}) + \frac{(X - \mathrm{i}Y)}{2\pi(1 + \kappa)} \cdot \frac{m(t)}{m'(t)t}$$
$$+ \kappa(A + \mathrm{i}B)Rt - (A - \mathrm{i}B)R\frac{m(t)}{m'(t)} - (C - \mathrm{i}D)\frac{R}{t} \qquad (5.23)$$

and for the stress problem

$$F_0(t) = \mathrm{i}\, R(t) + \frac{(X + \mathrm{i}Y)}{2\pi(1 + \kappa)}(\log t - \kappa \log \bar{t}) + \frac{(X - \mathrm{i}Y)}{2\pi(1 + \kappa)} \cdot \frac{m(t)}{\overline{m'(t)t}}$$

$$- (A + \mathrm{i}B)Rt - (A - \mathrm{i}B)R\frac{m(t)}{\overline{m'(t)}} - (C - \mathrm{i}D)\frac{R}{t} + \text{constant.} \tag{5.24}$$

Again as in Section 4.1 it may be confirmed that all functions in (5.22) are single valued in their regions of definition.

Other boundary conditions may also be referred to the image region. For example if we suppose the shear stress $\tau_{\xi\eta}$ and the normal displacement u_ξ are specified on C then from (5.14) and (5.16) the corresponding boundary conditions on $|\zeta| = 1$ are

$$\operatorname{Im}\left[-\frac{\bar{\zeta}}{\zeta m'(\zeta)}\left\{m(\zeta)\frac{\mathrm{d}}{\mathrm{d}\zeta}\overline{\left(\frac{\Omega'(\zeta)}{m'(\zeta)}\right)} + \overline{\omega'(\zeta)}\right\}\right] = \tau_{\xi\eta} \quad (\text{on } |\zeta| = 1) \tag{5.25}$$

and

$$\operatorname{Re}\left[\overline{\zeta\, m'(\zeta)}\left\{\kappa\, \Omega(\zeta) - \frac{m(\zeta)}{m'(\zeta)}\overline{\Omega'(\zeta)} - \overline{\omega(\zeta)}\right\}\right] = 2\mu|m'(\zeta)|u_\xi \quad (\text{on } |\zeta| = 1). \tag{5.26}$$

It will be noticed that the boundary conditions derived above are similar in form to those which hold in the corresponding problems of Chapter 4. Also, as the functions $\Omega(\zeta)$ and $\omega(\zeta)$ are holomorphic in either $|\zeta| \leqslant 1$ or $|\zeta| \geqslant 1$, the three basic methods of solution discussed in Chapter 4 may be applied to these problems. In the following sections we shall examine the direct method and the method based on continuation. As the series method will not be discussed in detail, its main features are indicated at this point.

To take a simple example, let us suppose we are dealing with an infinite region which is mapped onto the exterior E of the unit circle $|\zeta| = 1$ by the function

$$z = R\zeta + \frac{b_1}{\zeta} + \frac{b_2}{\zeta^2} + \cdots. \tag{5.27}$$

From (5.22) the stress and displacement boundary value problems reduce to the condition

$$\gamma\, \Omega_0(t) + \frac{m(t)}{m'(t)}\overline{\Omega_0'(t)} + \overline{\omega_0(t)} = F_0(t) \quad \text{on } t\bar{t} = 1. \tag{5.28}$$

By Laurent's theorem, since $\Omega_0(\zeta)$, $\omega_0(\zeta)$ are holomorphic and single valued in $|\zeta| \geqslant 1$ and bounded at infinity, they have expansions of the form

$$\Omega_0(\zeta) = \sum_{n=0}^{\infty} \alpha_n \zeta^{-n}, \qquad \omega_0(\zeta) = \sum_{n=0}^{\infty} \beta_n \zeta^{-n}. \tag{5.29}$$

Moreover, as $m'(\zeta) \neq 0$ for $|\zeta| \geqslant 1$, and the functions $F_0(t)$ and $m(t)/\overline{m'(t)}$ are differentiable and single valued on $t\bar{t} = 1$, they may be expanded in the Fourier series

$$\begin{aligned} F_0(t) &= \sum_{k=-\infty}^{\infty} F_k e^{ik\theta} = \sum_{k=-\infty}^{\infty} F_k t^k \\ \frac{m(t)}{\overline{m'(t)}} &= \sum_{k=-\infty}^{\infty} C_k t^k. \end{aligned} \tag{5.30}$$

On inserting the series expansions (5.29) and (5.30) into (5.28) and comparing coefficients of t^k, a set of simultaneous linear equations for the coefficients α_n, β_n are obtained. Simple solutions to these equations are only obtainable when the mapping function $m(\zeta)$ and the applied displacements or tractions have simple forms. We shall omit further discussion of this technique and the reader is referred to Muskhelishvili,[43] Part III and Sokolnikoff,[59] Chapter V, for more detailed calculations and illustrations. Some examples are given at the end of this chapter.

5.3 Direct method of solution

As in Section 4.3 we shall be concerned principally with the application of this method to the solution of the stress and displacement boundary value problems. For definiteness we shall suppose R_z is the region outside a hole with boundary C and is mapped onto the region $|\zeta| \geqslant 1$ by the function

$$z = m(\zeta) = R\zeta + \frac{m_1}{\zeta} + \frac{m_2}{\zeta^2} + \cdots + \frac{m_n}{\zeta^n} \tag{5.31}$$

where $m'(\zeta) \neq 0$ for $|\zeta| \geqslant 1$. The boundary conditions are, from (5.22)

$$\gamma\, \Omega_0(t) + \frac{m(t)}{\overline{m'(t)}}\, \overline{\Omega_0'(t)} + \overline{\omega_0(t)} = F_0(t) \tag{5.32}$$

where $\Omega_0(\zeta)$, $\omega_0(\zeta)$ are holomorphic and single valued in $|\zeta| \geqslant 1$ and $F_0(t)$ is single valued on $|\zeta| = 1$. On integrating through (5.32) with

$$\frac{1}{2\pi i} \int_\Gamma \frac{1}{t - \zeta}\, dt$$

where Γ is the contour $|t| = 1$ described in an anticlockwise manner, and we suppose $|\zeta| \geqslant 1$, we find

$$\frac{1}{2\pi i}\int_\Gamma \left(\gamma\, \Omega_0(t) + \frac{m(t)}{\overline{m'(t)}}\overline{\Omega_0'(t)} + \overline{\omega_0(t)}\right)\frac{dt}{(t-\zeta)} = \frac{1}{2\pi i}\int_\Gamma \frac{F_0(t)\,dt}{t-\zeta}.$$

Now as $\Omega_0(\zeta)$, $\omega_0(\zeta)$ are single-valued holomorphic functions for all $|\zeta| \geqslant 1$ including the point at infinity, the integral theorems of Section 1.5 imply

$$\begin{aligned} \frac{1}{2\pi i}\int_\Gamma \frac{\overline{\omega_0(t)}\,dt}{t-\zeta} &= 0 && (|\zeta| \geqslant 1) \\ \frac{1}{2\pi i}\int_\Gamma \frac{\Omega_0(t)\,dt}{t-\zeta} &= -\Omega_0(\zeta) + \Omega_0(\infty) && (|\zeta| \geqslant 1) \end{aligned} \qquad (5.33)$$

and hence

$$\gamma(\Omega_0(\zeta) - \Omega_0(\infty)) - \frac{1}{2\pi i}\int_\Gamma \frac{m(t)\,\overline{\Omega_0'(t)}\,dt}{\overline{m'(t)}\,(t-\zeta)} = -\frac{1}{2\pi i}\int_\Gamma \frac{F_0(t)\,dt}{t-\zeta}. \qquad (5.34)$$

Similarly, since the roles of $\Omega_0(\zeta)$ and $\omega_0(\zeta)$ may be interchanged in (5.33), $\omega_0(\zeta)$ is given by the equation

$$\omega_0(\zeta) - \omega_0(\infty) = -\frac{1}{2\pi i}\int_\Gamma \frac{\omega_0(t)\,dt}{t-\zeta} \quad (|\zeta| \geqslant 1)$$

and on substituting for $\omega_0(t)$ from the boundary condition (5.32)

$$\omega_0(\zeta) - \omega_0(\infty) = -\frac{1}{2\pi i}\int_\Gamma \frac{\overline{F_0(t)}\,dt}{t-\zeta} + \frac{1}{2\pi i}\int_\Gamma \frac{\overline{m(t)}\,\Omega_0'(t)\,dt}{m'(t)\,(t-\zeta)} \quad (|\zeta| \geqslant 1). \qquad (5.35)$$

The solution to the problem may be completed when the integrals involving $\Omega_0'(t)$ in (5.34) and (5.35) have been evaluated.

To evaluate the unknown integral in (5.34) we show that it is possible to define a function which is holomorphic in $|\zeta| < 1$, except at a finite number of points where simple poles exist, and which has the boundary value $m(t)\,\overline{\Omega_0'(t)}/\overline{m'(t)}$ on $|\zeta| = 1$. Since on $t\bar{t} = 1$

$$\frac{m(t)\,\overline{\Omega_0'(t)}}{\overline{m'(t)}} = \frac{m(t)\,\overline{\Omega_0'(1/\bar{t})}}{\overline{m'(1/\bar{t})}}$$

the required function would seem to be

$$m(\zeta)\,\overline{\Omega_0'(1/\bar{\zeta})}/\overline{m'(1/\bar{\zeta})} \qquad (5.36)$$

for $|\zeta| < 1$, if we suppose $m(\zeta)$ is defined by (5.31) for all ζ. The singularities of (5.36) must be found before the integral in (5.34) may be evaluated. By definition $m(\zeta)$ has a pole of order n at $\zeta = 0$. Also since $m'(\zeta) \neq 0$ for all $|\zeta| \geqslant 1$

$$\overline{m'(1/\bar{\zeta})} = R - \bar{m}_1\zeta^2 - \cdots - nm_n\zeta^{n+1} \neq 0 \quad \text{for } |\zeta| \leqslant 1.$$

Thus the function $m(\zeta)/\overline{m'(1/\bar{\zeta})}$ is holomorphic for $|\zeta| \leqslant 1$ except at $\zeta = 0$ where there exists a pole of order n. Further, since the function $\overline{\Omega_0'(1/\bar{\zeta})}$ is holomorphic for $|\zeta| \leqslant 1$ and $\Omega_0(\zeta)$ has the expansion

$$\Omega_0(\zeta) = \alpha_0 + \frac{\alpha_1}{\zeta} + \frac{\alpha_2}{\zeta^2} + \cdots \quad \text{for large } |\zeta| \tag{5.37}$$

then

$$\overline{\Omega_0'(1/\bar{\zeta})} = -\bar{\alpha}_1\zeta^2 - 2\bar{\alpha}_2\zeta^3 - \cdots \quad \text{for small } |\zeta| \tag{5.38}$$

and the function $m(\zeta)\,\overline{\Omega_0'(1/\bar{\zeta})}/\overline{m'(1/\bar{\zeta})}$ has a pole of order $n - 2$ at $\zeta = 0$. We shall assume for the present that $n > 2$. To evaluate the integral in (5.34) we require the expansion of this function about the origin, namely

$$\frac{m(\zeta)}{\overline{m'(1/\bar{\zeta})}}\,\overline{\Omega_0'(1/\bar{\zeta})} = \frac{\kappa_{n-2}}{\zeta^{n-2}} + \frac{\kappa_{n-3}}{\zeta^{n-3}} + \cdots + \frac{\kappa_1}{\zeta} + \kappa_0 + \cdots. \tag{5.39}$$

If $m(\zeta)/\overline{m'(1/\bar{\zeta})}$ has a principal part of the form

$$\frac{m(\zeta)}{\overline{m'(1/\bar{\zeta})}} = \frac{b_n}{\zeta^n} + \frac{b_{n-1}}{\zeta^{n-1}} + \cdots + \frac{b_1}{\zeta} + b_0 + \cdots \quad \text{for small } |\zeta| \tag{5.40}$$

then the κ_i in (5.39) are simply found by multiplication of the series in (5.38) and (5.40) to be

$$\begin{aligned} \kappa_{n-2} &= -b_n\bar{\alpha}_1 \\ \kappa_{n-3} &= -b_{n-1}\bar{\alpha}_1 - b_n 2\bar{\alpha}_2 \\ &\cdots\cdots\cdots \\ \kappa_1 &= -b_3\bar{\alpha}_1 - b_4 2\bar{\alpha}_2 - \cdots - b_n(n-2)\bar{\alpha}_{n-2} \\ \kappa_0 &= -b_2\bar{\alpha}_1 - b_3 2\bar{\alpha}_2 - \cdots - b_n(n-1)\bar{\alpha}_{n-1}. \end{aligned} \tag{5.41}$$

On returning to the integral in (5.34) we find

$$\frac{1}{2\pi i}\int_\Gamma \frac{m(t)\,\overline{\Omega_0'(t)}\,dt}{\overline{m'(t)}(t-\zeta)} = \frac{1}{2\pi i}\int_{\Gamma_0} \frac{m(\tau)\,\overline{\Omega_0'(1/\bar{\tau})}\,d\tau}{\overline{m'(1/\bar{\tau})}(\tau-\zeta)} \quad |\zeta| \geqslant 1$$

where Γ_0 is a circle of small radius about $\zeta = 0$, and hence from (5.39) this integral has the value

$$-\frac{\kappa_{n-2}}{\zeta^{n-2}} - \frac{\kappa_{n-3}}{\zeta^{n-3}} - \cdots - \frac{\kappa_1}{\zeta}.$$

Thus

$$\Omega_0(\zeta) - \alpha_0 + \frac{1}{\gamma}\left(\frac{\kappa_1}{\zeta} + \cdots + \frac{\kappa_{n-2}}{\zeta^{n-2}}\right) = -\frac{1}{2\pi i\gamma}\int_\Gamma \frac{F_0(t)\,dt}{t-\zeta} \quad (|\zeta| \geqslant 1). \tag{5.42}$$

Further relations satisfied by the constants $\kappa_1, \kappa_2, \ldots, \kappa_{n-2}$ may be found on expanding (5.42) in series in $1/\zeta$ for large $|\zeta|$, giving, from (5.37)

$$\begin{aligned} \gamma\alpha_k + \kappa_k &= F_k \quad \text{for } k = 1, 2, \ldots, n-2 \\ \gamma\alpha_s &= F_s \quad \text{for } s = n-1, n, \ldots \end{aligned} \tag{5.43}$$

where

$$F_k = \frac{1}{2\pi i}\int_\Gamma F_0(t)t^{k-1}\,dt. \tag{5.44}$$

On substituting in (5.41) for κ_k, $k = 1, 2, \ldots, n-2$, a set of $n-2$ linear equations is obtained from which the $n-2$ constants $\alpha_1, \alpha_2, \ldots, \alpha_{n-2}$ may be found.† Hence the κ_k are determined from (5.43) and $\Omega_0(\zeta)$ is given by (5.42) in terms of the arbitrary constant α_0.

A similar approach may be employed to determine the integral involving $\Omega_0'(t)$ in (5.35). In this case it is clear that

$$\begin{aligned} \int_\Gamma \frac{\overline{m(t)}\,\Omega_0'(t)\,dt}{m'(t)(t-\zeta)} &= \int_{\Gamma_\infty} \frac{\overline{m(1/\bar{\tau})}\,\Omega_0'(\tau)\,d\tau}{m'(\tau)(\tau-\zeta)} \\ &\quad -2\pi i\frac{\overline{m(1/\bar{\zeta})}}{m'(\zeta)}\,\Omega_0'(\zeta) \quad (|\zeta| \geqslant 1) \end{aligned}$$

where Γ_∞ is a circle of large radius about the origin described in an anti-clockwise manner, and the last term above results from the simple pole at the point ζ. Using the definitions (5.40) and (5.37) it will be seen that

$$\frac{\overline{m(1/\bar{\zeta})}}{m'(\zeta)} = \bar{b}_n\zeta^n + \bar{b}_{n-1}\zeta^{n-1} + \cdots + \bar{b}_1\zeta + \bar{b}_0 + \cdots \quad \text{for large } |\zeta|$$

and hence

$$\begin{aligned} \frac{\overline{m(1/\bar{\zeta})}}{m'(\zeta)}\,\Omega_0'(\zeta) &= (\bar{b}_n\zeta^n + \cdots + \bar{b}_1\zeta + \bar{b}_0 + \cdots)\left(-\frac{\alpha_1}{\zeta^2} - \cdots - \frac{n\alpha_n}{\zeta^{n+1}} - \cdots\right) \\ &= \bar{\kappa}_{n-2}\zeta^{n-2} + \bar{\kappa}_{n-3}\zeta^{n-3} + \cdots + \bar{\kappa}_1\zeta + \bar{\kappa}_0 + \cdots \quad \text{for large } |\zeta| \end{aligned} \tag{5.45}$$

† Because of the form of (5.41) it is more convenient to determine the α_k and then κ_k rather than the κ_k directly.

from (5.41). Thus

$$\frac{1}{2\pi i}\int_{\Gamma_\infty} \frac{\overline{m(1/\bar{\zeta})}\,\Omega_0'(\tau)\,d\tau}{m'(\tau)(\tau-\zeta)} = \bar{\kappa}_{n-2}\zeta^{n-2} + \cdots + \bar{\kappa}_1\zeta + \bar{\kappa}_0.$$

Hence $\omega_0(\zeta)$ is defined by the equation

$$\omega_0(\zeta) - \omega_0(\infty) = -\frac{1}{2\pi i}\int_\Gamma \frac{\overline{F_0(t)}\,dt}{t-\zeta} - \frac{\overline{m(1/\bar{\zeta})}}{m'(\zeta)}\Omega_0'(\zeta) + \bar{\kappa}_0 + \bar{\kappa}_1\zeta + \cdots + \bar{\kappa}_{n-2}\zeta^{n-2}. \tag{5.46}$$

It may be confirmed that $\omega_0(\zeta)$ is holomorphic at infinity in view of (5.45).

The above calculations are clearly only valid for $n > 2$, since for $n \leqslant 2$ the functions defined in (5.39) and (5.45) are holomorphic at the origin and infinity respectively. In these cases equation (5.42) becomes

$$\Omega_0(\zeta) - \Omega_0(\infty) = -\frac{1}{2\pi i\gamma}\int_\Gamma \frac{F_0(t)\,dt}{t-\zeta} \quad (|\zeta| \geqslant 1) \tag{5.47}$$

and from (5.46)

$$\omega_0(\zeta) - \omega_0(\infty) = -\frac{1}{2\pi i}\int_\Gamma \frac{\overline{F_0(t)}\,dt}{t-\zeta} - \frac{\overline{m(1/\bar{\zeta})}}{m'(\zeta)}\Omega_0'(\zeta) + \kappa_0 \tag{5.48}$$

where

$$\kappa_0 = \lim_{|\zeta|\to\infty} \frac{\overline{m(1/\bar{\zeta})}}{m'(\zeta)}\Omega_0'(\zeta) \quad (=0 \quad \text{for } n < 2). \tag{5.49}$$

The discussion following equation (5.35) has been restricted by the assumption that the transformation $m(\zeta)$ was of the form (5.31) having a pole of order n at the origin. Although by Laurent's theorem, every transformation $m(\zeta)$ has a representation of the form (5.31) as $n \to \infty$, even a simple rational function may require a large number of terms to achieve a solution of sufficient accuracy. In fact the solution given above may be extended on assuming $m(\zeta)$ is a rational function with several poles lying in the region $|\zeta| < 1$ and such that $m'(\zeta) \neq 0$ for $|\zeta| \geqslant 1$. Again the integrals in (5.34) and (5.35) may be evaluated by residue theory, although in the process further sets of constants equivalent to the κ_i of (5.39) are generated at each pole. These constants are found to satisfy certain systems of linear equations and hence the complex potentials may be determined. Further details are given in the next section and by Muskhelishvili,[43] Section 85.

1. *Unstressed elliptical hole*

As a first illustration we examine the case of an unstressed elliptical hole in an infinite plate under a uniform tension at infinity. It will be seen from (5.5) that the region may be mapped onto the exterior of the unit circle $|\zeta| \geqslant 1$ by

$$z = R(\zeta + m\zeta^{-1}) \quad (0 \leqslant m < 1) \tag{5.50}$$

where R and m are related to the semi-axes a, b of the ellipse by

$$R(1 + m) = a, \qquad R(1 - m) = b. \tag{5.51}$$

Assuming the uniform tension T at infinity acts at an angle α to the major axis of the ellipse then in (5.18)

$$A = \tfrac{1}{4}T, \qquad C + \mathrm{i}D = -\tfrac{1}{2}T\mathrm{e}^{-2\mathrm{i}\alpha}.$$

If the hole is unstressed and the rotation at infinity is zero then in (5.24)

$$F_0(t) = -\frac{RT}{2}t - \frac{RT}{4}\left(\frac{t^2 + m}{1 - mt^2}\right)\frac{1}{t} + \frac{RT\mathrm{e}^{2\mathrm{i}\alpha}}{2t}.$$

The solution may now be calculated from (5.47) and (5.48). In (5.47) on evaluating the residues of the pole at the origin we find

$$\Omega_0(\zeta) - \Omega_0(\infty) = \frac{RT}{4\zeta}(2\mathrm{e}^{2\mathrm{i}\alpha} - m) \tag{5.52}$$

and similarly

$$\omega_0(\zeta) - \omega_0(\infty)$$
$$= -\frac{RT}{4\zeta} - \frac{RT}{4}\cdot\frac{(m^2 + 1)\zeta}{(\zeta^2 - m)} + \frac{RT}{4}\cdot\frac{(2\mathrm{e}^{2\mathrm{i}\alpha} - m)(1 + m\zeta^2)}{(\zeta^2 - m)\zeta}. \tag{5.53}$$

As we are considering a stress boundary value problem the constant terms $\Omega_0(\infty)$, $\omega_0(\infty)$ in (5.52) and (5.53) may be ignored and the total potentials are given from (5.18) by

$$\begin{aligned}\Omega(\zeta) &= \tfrac{1}{4}RT\zeta + \Omega_0(\zeta)\\ \omega(\zeta) &= -\tfrac{1}{2}RT\mathrm{e}^{-2\mathrm{i}\alpha}\zeta + \omega_0(\zeta).\end{aligned} \tag{5.54}$$

These results will be examined in more detail in Section 5.5 where this problem is solved by means of continuation.

2. *Rigid elliptical inclusion*

It should be noticed that the above method also allows us to determine the solutions to displacement boundary value problems. As an example let us consider an infinite medium bonded to a rigid elliptical inclusion and suppose the medium is deformed by giving the inclusion a rotation through a small angle ε, the stresses and rotation at infinity being zero. Then on the elliptical boundary of the medium the displacement is specified to be

$$D = \mathrm{i}\varepsilon z$$

so that on transforming to the region $|\zeta| \geqslant 1$ by $z = R(\zeta + m\zeta^{-1})$ the boundary conditions take the form of (5.22) where $\gamma = -\kappa$ and from (5.23)

$$F_0(t) = -2\mu\, D(t) = -2\mu\mathrm{i}\varepsilon R(t + mt^{-1})$$

there being a zero resultant force applied to the inclusion. Hence on using the solution (5.47) and (5.48) derived above we find

$$\begin{aligned}\Omega_0(\zeta) - \Omega_0(\infty) &= -\frac{\mu\varepsilon R}{\pi\kappa}\int_\Gamma \left(t + \frac{m}{t}\right)\frac{\mathrm{d}t}{(t-\zeta)} \\ &= \frac{2\mu\varepsilon Rm\mathrm{i}}{\kappa\zeta} \qquad (5.55)\end{aligned}$$

$$\omega_0(\zeta) - \omega_0(\infty) = \frac{2\mu\varepsilon R\mathrm{i}}{\zeta}\left(1 + \frac{m}{\kappa}\left(\frac{1 + m\zeta^2}{\zeta^2 - m}\right)\right). \qquad (5.56)$$

As the terms $\Omega_0(\infty)$, $\omega_0(\infty)$ merely contribute to a rigid-body translation they may be ignored. It will be seen from (5.18) that $\Omega_0(\zeta)$, $\omega_0(\zeta)$ are the complete potentials for the solution.

It is of interest to calculate the moment M of the couple necessary to produce the rotation through the angle ε. From equation (2.48)

$$M = -\mathrm{Re}\left[z\bar{z}\,\Omega'(z) + z\,\omega(z) - \int \omega(z)\,\mathrm{d}z\right]_C$$

where the circuit around the ellipse C is taken in a clockwise direction. As $\Omega'(z)$, $\omega(z)$ are single-valued functions

$$M = \mathrm{Re}\left[\int \omega(\zeta)m'(\zeta)\,\mathrm{d}\zeta\right]_C$$

which is simply the contour integral

$$M = -\mathrm{Re}\left[\int_\Gamma \frac{2\mu\varepsilon R\mathrm{i}}{\zeta}\left\{1 + \frac{m}{\kappa}\left(\frac{1 + m\zeta^2}{\zeta^2 - m}\right)\right\} R\left(1 - \frac{m}{\zeta^2}\right)\mathrm{d}\zeta\right].$$

Thus on evaluating the residue at the origin

$$M = 4\pi\mu R^2\left(1 + \frac{m^2}{\kappa}\right)\varepsilon. \qquad (5.57)$$

3. *Hypotrochoidal hole*

As a more general illustration we consider an infinite plane containing a hypotrochoidal hole which is inflated by a uniform pressure p, the stress and rotation at infinity being zero. Since from (5.6) the exterior of the hole is mapped conformally onto $|\zeta| \geqslant 1$ by

$$z = R\left(\zeta + \frac{m}{\zeta^n}\right) \quad \left(n > 2, \quad 0 \leqslant m < \frac{1}{n}\right)$$

the complex potentials satisfy the boundary condition (5.22) where from Section 2.12 the resultant force is given by $R(z) = \mathrm{i}pz$ so that in (5.24)

$$F_0(t) = -pR\left(t + \frac{m}{t^n}\right). \tag{5.58}$$

To determine the solution from (5.42) and (5.46) we require the values of the constants $\kappa_0, \kappa_1, \ldots, \kappa_{n-2}$.

From (5.40)

$$\frac{m(\zeta)}{\overline{m'(1/\bar{\zeta})}} = \left(\zeta + \frac{m}{\zeta^n}\right)(1 - nm\zeta^{n+1})^{-1} = \frac{m}{\zeta^n} + O(\zeta) \quad \text{for small } |\zeta|$$

thus

$$b_n = m$$
$$b_k = 0 \quad \text{for } k = 1, 2, \ldots, n - 1.$$

and hence equations (5.41) become

$$\kappa_{n-1-k} = -mk\bar{\alpha}_k \quad \text{for } k = 1, 2, \ldots, n - 1.$$

Prior to substituting for κ_k from (5.43) we should evaluate the integrals

$$F_k = \frac{1}{2\pi\mathrm{i}} \int_\Gamma F_0(t)t^{k-1}\,\mathrm{d}t = -\frac{pR}{2\pi\mathrm{i}} \int_\Gamma \left(t + \frac{m}{t^n}\right)t^{k-1}\,\mathrm{d}t$$
$$= \begin{cases} 0 & (k \neq n) \\ -pRm & (k = n). \end{cases}$$

Hence $\kappa_k = -\alpha_k$ for $k = 1, 2, \ldots, n - 2$ so that the α_k satisfy the equations

$$\alpha_{n-1-k} = mk\bar{\alpha}_k \quad (k = 1, 2, \ldots, n - 2)$$

and

$$\alpha_n = -pRm.$$

These relations imply

$$\alpha_k = m(n - 1 - k)\bar{\alpha}_{n-1-k} = m^2(n - 1 - k)k\alpha_k$$

so that $\alpha_k = 0$ unless $k(n - 1 - k) = m^{-2}$ for some k in the range $k = 1, 2, \ldots, n - 2$. This possibility does not arise as we have assumed $mn < 1$. Thus $\kappa_k = 0$ for $k = 1, 2, \ldots, n - 2$, and the complex potentials follow from (5.42) and (5.46)

$$\begin{aligned} \Omega(\zeta) &= \alpha_0 + \frac{pR}{2\pi i}\int_\Gamma \left(t + \frac{m}{t^n}\right)\frac{dt}{t - \zeta} \\ &= \alpha_0 - \frac{pRm}{\zeta^n} \quad (|\zeta| \geqslant 1, \quad n > 2) \end{aligned} \tag{5.59}$$

and

$$\omega(\zeta) = \omega_0(\infty) + \bar{\kappa}_0 - \frac{pR}{\zeta} - \frac{pRmn}{\zeta}\cdot\frac{(1 + m\zeta^{n+1})}{(\zeta^{n+1} - nm)} \quad (|\zeta| \geqslant 1, \quad n > 2). \tag{5.60}$$

The constants in $\Omega(\zeta)$ and $\omega(\zeta)$ may be omitted since they only contribute to a rigid-body motion. The stresses and displacements in the medium may now be evaluated from equations (5.14) and (5.15).

5.4 Stress and displacement continuations

Let us suppose that the mapping $z = m(\zeta)$ takes the region R_z of the z-plane occupied by the elastic body into the region S^+ of the ζ-plane. Assuming S^+ has as a boundary the unit circle $|\zeta| = 1$, we denote the inverse region of S^+ in $|\zeta| = 1$ by S^- so that, for example, if S^+ is the region $|\zeta| > 1$ then S^- is $|\zeta| < 1$.

As in the earlier chapters (Sections 3.3, 4.4) the stress continuation of the complex potentials may be defined by considering the stress boundary value problem and assuming the applied stress is zero. Similarly on assuming the applied displacement distribution is zero in the displacement boundary value problem the form of the displacement continuation results. In both cases the complex potentials satisfy the boundary conditions (5.21)

$$\gamma\,\Omega^+(t) + \frac{m^+(t)}{m'^+(t)}\,\overline{\Omega'^+(t)} + \overline{\omega^+(t)} = 0 \quad \text{on } t\bar{t} = 1 \tag{5.61}$$

where $\gamma = 1$ for the stress problem, $\gamma = -\kappa$ for the displacement problem and the suffix $^+$ denotes the limiting value attained by a function defined in S^+ on the boundary $t\bar{t} = 1$. Now proceeding as in Section 4.4 the boundary conditions (5.61) may be written as†

$$\Omega^+(t) = -\frac{1}{\gamma}\left\{\frac{m(t)}{\overline{m'^-(1/\bar{t})}}\,\overline{\Omega'^-(1/\bar{t})} + \overline{\omega^-(1/\bar{t})}\right\}$$

† Compare with equation (4.44).

which expresses the fact that the holomorphic function $\Omega(\zeta)$ for $\zeta \in S^+$ and the function

$$-\frac{1}{\gamma}\left\{\frac{m(\zeta)}{\overline{m'(1/\bar{\zeta})}}\overline{\Omega'(1/\bar{\zeta})} + \overline{\omega(1/\bar{\zeta})}\right\}$$

which is holomorphic for $\zeta \in S^-$ (except at certain extreme points which will be examined separately) are continuous across $|\zeta| = 1$. We therefore extend the definition of $\Omega(\zeta)$ from S^+ into S^- by

$$\Omega(\zeta) = -\frac{1}{\gamma}\left\{\frac{m(\zeta)}{\overline{m'(1/\bar{\zeta})}}\overline{\Omega'(1/\bar{\zeta})} + \overline{\omega(1/\bar{\zeta})}\right\} \quad \text{for } \zeta \in S^-. \tag{5.62}$$

On replacing ζ by $1/\bar{\zeta}$ in this equation and taking the conjugate

$$\omega(\zeta) = -\gamma\,\overline{\Omega(1/\bar{\zeta})} - \frac{\overline{m(1/\bar{\zeta})}}{m'(\zeta)}\Omega'(\zeta) \quad (\zeta \in S^+). \tag{5.63}$$

It should be noticed that as $m(\zeta)$ is only defined for $\zeta \in S^+$ we can only deduce (5.62) from (5.61) if we define $m(\zeta)$ for $\zeta \in S^-$ to be an analytic continuation of $m(\zeta)$ for $\zeta \in S^+$ across the boundary $|\zeta| = 1$. When $m(\zeta)$ is a polynomial or a rational function it has the same functional form in both S^+ and S^-, and we shall assume $m(\zeta)$ is a rational function in the following.

Again we shall largely confine our attention to the stress continuation which occurs when $\gamma = 1$. Then from (5.62) and (5.63) $\Omega(\zeta)$ is continued into S^- by

$$\Omega(\zeta) = -\frac{m(\zeta)}{\overline{m'(1/\bar{\zeta})}}\overline{\Omega'(1/\bar{\zeta})} - \overline{\omega(1/\bar{\zeta})} \quad (\zeta \in S^-) \tag{5.64}$$

and $\omega(\zeta)$ is defined in terms of $\Omega(\zeta)$ by

$$\omega(\zeta) = -\overline{\Omega(1/\bar{\zeta})} - \frac{\overline{m(1/\bar{\zeta})}}{m'(\zeta)}\Omega'(\zeta) \quad (\zeta \in S^+). \tag{5.65}$$

Once again this could be regarded as a useful definition of $\omega(\zeta)$, $\zeta \in S^+$ in terms of $\Omega(\zeta)$ in S^+ and S^-, where the values of $\Omega(\zeta)$ in these two regions are quite independent. On substituting for $\omega(\zeta)$ from (5.65) into the stress and displacement relations (5.14) and (5.15) we find

$$2\mu D = \kappa\,\Omega(\zeta) + \Omega(1/\bar{\zeta}) - (m(\zeta) - m(1/\bar{\zeta}))\frac{\overline{\Omega'(\zeta)}}{\overline{m'(\zeta)}} \tag{5.66}$$

$$\tau_{\xi\xi} + \mathrm{i}\tau_{\xi\eta} = W(\zeta) - \frac{m'(1/\bar{\zeta})}{\zeta\bar{\zeta}\,\overline{m'(\zeta)}}W(1/\bar{\zeta}) + \left(1 - \frac{m'(1/\bar{\zeta})}{\zeta\bar{\zeta}\,\overline{m'(\zeta)}}\right)\overline{W(\zeta)}$$
$$+ (m(1/\bar{\zeta}) - m(\zeta))\frac{\overline{\zeta\,W'(\zeta)}}{\zeta\,\overline{m'(\zeta)}} \tag{5.67}$$

where

$$W(\zeta) = \Omega'(\zeta)/m'(\zeta). \tag{5.68}$$

Hence on the boundary $|\zeta| = 1$ the stress continuation leads to the relations

$$2\mu D = \kappa\, \Omega^+(t) + \Omega^-(t)$$

$$\tau_{\xi\xi} + \mathrm{i}\tau_{\xi\eta} = W^+(t) - W^-(t) = \frac{1}{m'(t)}(\Omega'^+(t) - \Omega'^-(t)). \tag{5.69}$$

Thus the stress continuation (5.64) has been chosen so that $\Omega'(\zeta)$ is continuous across those parts of the boundary on which $\tau_{\xi\xi} + \mathrm{i}\tau_{\xi\eta} = 0$.

We must now investigate the requirements imposed on $\Omega(\zeta)$, for $\zeta \in S^-$, by the condition that $\Omega(\zeta)$ and $\omega(\zeta)$ are holomorphic for $\zeta \in S^+$ and have a known behaviour at infinity if S^+ is the region $|\zeta| > 1$.

First let us suppose R_z is a bounded simply connected region mapped onto the interior of the unit circle, so that S^+ is $|\zeta| < 1$, and S^-, $|\zeta| > 1$. Since $m(\zeta)$ is a rational function, holomorphic in $|\zeta| < 1$, it may have a finite number of poles in $|\zeta| > 1$, say at $\zeta_1, \zeta_2, \ldots, \zeta_r$ with orders $m_1, m_2, \ldots, m_r$ respectively and at infinity with order m_∞. From (5.64) since $\overline{\Omega'(1/\bar{\zeta})}$, $\overline{\omega(1/\bar{\zeta})}$ are holomorphic functions for all points $\zeta \in S^-$, and $\overline{m'(1/\bar{\zeta})} \neq 0$ for $|\zeta| > 1$, then $\Omega(\zeta)$ is holomorphic for $\zeta \in S^-$ except at the points where $m(\zeta)$ has a pole. In fact the singularities of $\Omega(\zeta)$ correspond exactly with those of $m(\zeta)$ in position and order, so that $\Omega'(\zeta)$ has poles at the points $\zeta_1, \zeta_2, \ldots, \zeta_r$ of orders $m_1 + 1, m_2 + 1, \ldots, m_r + 1$, respectively and at infinity of order $m_\infty - 1$.

Secondly let us suppose R_z is an infinite region which maps onto the exterior of the unit circle $|\zeta| > 1$. Then in this case S^+ and S^- are the regions $|\zeta| > 1$ and $|\zeta| < 1$ respectively. Now as the mapping has the form (5.4) namely

$$z = R\zeta + O(\zeta^{-1}) \quad \text{for large } |\zeta|$$

the potentials $\Omega(\zeta)$, $\omega(\zeta)$ have the following behaviour for large $|\zeta|$ (see (5.18))

$$\begin{aligned} \Omega(\zeta) &= (A + \mathrm{i}B)R\zeta - \frac{(X + \mathrm{i}Y)}{2\pi(1+\kappa)}\log\zeta + O(1) \\ \omega(\zeta) &= (C + \mathrm{i}D)R\zeta + \frac{\kappa(X - \mathrm{i}Y)}{2\pi(1+\kappa)}\log\zeta + O(1) \end{aligned} \tag{5.70}$$

and hence we see that the function $\overline{\Omega'(1/\bar{\zeta})}$ is holomorphic for $|\zeta| \leqslant 1$ even at $\zeta = 0$, but that $\overline{\omega(1/\bar{\zeta})}$ is only holomorphic in $|\zeta| \leqslant 1$ for $\zeta \neq 0$ since

$$\overline{\omega(1/\bar{\zeta})} = (C - \mathrm{i}D)\frac{R}{\zeta} - \frac{\kappa(X + \mathrm{i}Y)}{2\pi(1+\kappa)}\log\zeta + O(1) \quad \text{for small } |\zeta|. \tag{5.71}$$

Again since $m(\zeta)$ is assumed to be a rational function, holomorphic and with $m'(\zeta) \neq 0$ in $|\zeta| \geqslant 1$, it may have a set of poles in $|\zeta| < 1$. Thus from (5.64) the singularities of $\Omega(\zeta)$ in $|\zeta| < 1$ are exactly those of $m(\zeta)$ together with a term of the form (5.71) at the origin. Thus $\Omega'(\zeta)$ has poles in $|\zeta| < 1$ corresponding to those of $m'(\zeta)$ in position and order together with a pole of order 2 at the origin, which has as its principal part†

$$(C - iD)\frac{R}{\zeta^2} + \frac{\kappa(X + iY)}{2\pi(1 + \kappa)}\frac{1}{\zeta} \tag{5.72}$$

and is holomorphic elsewhere in $|\zeta| \leqslant 1$.

It should be noted in the above discussions that we have tacitly assumed $\Omega'(\zeta)$, $\omega'(\zeta)$ are holomorphic in S^+, thereby removing the possibility of point forces or moments being present in the bodies. It is a simple matter to extend the above analysis to cope with such singularities and this is left to the reader.

In the solutions to the various boundary value problems which follow from this continuation arbitrary functions arise which have the same distribution of poles as $\Omega'(\zeta)$. In order to determine these functions we require a full knowledge of the principal part of $\Omega'(\zeta)$ at each of its poles which must be calculated from (5.64). It turns out to be less complicated to determine the arbitrary functions by making the equivalent requirement that $\omega(\zeta)$ in (5.65) is holomorphic for $\zeta \in S^+$ (and has the correct behaviour for large $|\zeta|$ if S^+ is the region $|\zeta| \geqslant 1$). These conditions yield a sufficient number of equations to determine the unknown constants in $\Omega(\zeta)$ (apart from those corresponding to a rigid-body motion in the case of a stress boundary value problem, which implies $\Omega(\zeta)$ is undetermined to within a term of the form $\alpha + iA\, m(\zeta)$ where A is real). Algebraically it proves convenient to differentiate (5.65) and require

$$\omega'(\zeta) = \frac{1}{\zeta^2}\overline{\Omega'(1/\bar{\zeta})} - \frac{d}{d\zeta}\left\{\frac{\overline{m(1/\bar{\zeta})}}{m'(\zeta)}\Omega'(\zeta)\right\} \tag{5.73}$$

is holomorphic for $\zeta \in S^+$ and if necessary has the correct behaviour at infinity from (5.70). We shall follow Milne-Thomson[41] and refer to this as the *holomorphy condition.*

5.5 Stress and displacement problems

Let us suppose the normal and shear stresses $\tau_{\xi\xi} + i\tau_{\xi\eta}$ are specified on the boundary C of the region R_z. Then on mapping R_z into the region S^+

† It should be remarked that if $m(\zeta)$ has a pole at the origin, the principal part will be different from (5.72). The terms of (5.72) are merely the contribution to this pole due to the behaviour of $\omega'(\zeta)$ at infinity.

of the ζ-plane by $z = m(\zeta)$ we find

$$\tau_{\xi\xi} + \mathrm{i}\tau_{\xi\eta} = f(z) = f(m(t)) \equiv f(t) \quad \text{on } |t| = 1.$$

On using the stress continuation (5.64), the stress and displacement fields may be represented by (5.66) and (5.67) in terms of a single potential $\Omega(\zeta)$ holomorphic in S^+ and S^- (except at certain points) and which is such that the boundary conditions become

$$\frac{1}{m'(t)}\{\Omega'^{+}(t) - \Omega'^{-}(t)\} = f(t) \quad \text{on } |t| = 1.$$

This is a simple Cauchy problem of a form already considered for a circular boundary. The difference with Section 4.5 lies in the fact that in this instance $\Omega'(\zeta)$ may have poles in S^-. Hence the solution takes the form

$$\Omega'(\zeta) = \frac{1}{2\pi\mathrm{i}}\int_\Gamma \frac{f(t)m'(t)\,\mathrm{d}t}{t - \zeta} + \psi(\zeta) \qquad (5.74)$$

where Γ is the contour $|t| = 1$ described so that S^+ lies to the left. As the Cauchy integral in (5.74) is holomorphic in S^+ and S^- the arbitrary function $\psi(\zeta)$ must have exactly the same singularities as $\Omega'(\zeta)$ and be holomorphic elsewhere in the whole plane. Hence the form of $\psi(\zeta)$ may be written down as a sum of principal parts of the poles, since their position and order are known from Section 5.4. The unknown constants occurring in this expression may be determined by substituting $\Omega'(\zeta)$ from (5.74) into the holomorphy condition (5.73), and hence the solution may be completed. In many cases this method proves easier than the direct method outlined in Section 5.3.

1. *Inflation of a hypotrochoidal hole*

As a representative illustration we consider the inflation of a hypotrochoidal hole in an infinite plate by a uniform pressure $\tau_{\xi\xi} + \mathrm{i}\tau_{\xi\eta} = -p$ with zero stress and rotation at infinity. If the plate is conformally mapped onto the region $|\zeta| \geqslant 1$ by

$$z = m(\zeta) = R(\zeta + m\zeta^{-n}) \quad (0 \leqslant m < 1/n)$$

then $m(\zeta)$ has a pole of order n at $\zeta = 0$. To avoid particular cases we shall assume $n > 2$. Then $\Omega'(\zeta)$ has a pole of order $n + 1$ at $\zeta = 0$ and from (5.70) behaves as $O(\zeta^{-2})$ for large $|\zeta|$. Thus from (5.74) $\psi(\zeta)$ must have the form

$$\frac{A_{n+1}}{\zeta^{n+1}} + \cdots + \frac{A_1}{\zeta} + A_0$$

corresponding to the pole at $\zeta = 0$ and hence

$$\Omega'(\zeta) = \frac{1}{2\pi i}\int_\Gamma -pR\left(1 - \frac{mn}{t^{n+1}}\right)\frac{dt}{t-\zeta} + \frac{A_{n+1}}{\zeta^{n+1}} + \cdots + \frac{A_1}{\zeta} + A_0. \quad (5.75)$$

As

$$\frac{1}{2\pi i}\int_\Gamma \left(1 - \frac{mn}{t^{n+1}}\right)\frac{dt}{t-\zeta}\begin{cases} = -1 & (|\zeta| < 1) \\ = -\dfrac{mn}{\zeta^{n+1}} & (|\zeta| > 1) \end{cases} \quad (5.76)$$

Γ being described in a clockwise manner, and $\Omega'(\zeta) = O(\zeta^{-2})$ for large $|\zeta|$, $A_0 = A_1 = 0$. The remaining constants $A_2, \ldots, A_{n+1}$ follow on substituting $\Omega'(\zeta)$ from (5.75) into the holomorphy condition (5.73). In this case

$$\omega'(\zeta) = \frac{1}{\zeta^2}(pR + \bar{A}_{n+1}\zeta^{n+1} + \cdots + \bar{A}_2\zeta^2)$$
$$- \frac{d}{d\zeta}\left\{\frac{(1 + m\zeta^{n+1})\zeta^n}{(\zeta^{n+1} - mn)}\left(\frac{A_{n+1} + pRmn}{\zeta^{n+1}} + \frac{A_n}{\zeta^n} + \cdots + \frac{A_2}{\zeta^2}\right)\right\}. \quad (5.77)$$

This function must be holomorphic for $|\zeta| \geqslant 1$ and have $O(\zeta^{-2})$ for large $|\zeta|$. These conditions imply

$$\bar{A}_{n-k+1} - (n-k)A_k = 0 \quad (k = 2, 3, \ldots, n-1)$$
$$A_{n+1} = A_n = 0$$

so that

$$A_k = 0 \quad (k = 2, 3, \ldots, n+1).$$

Hence we have the very simple solution

$$\begin{aligned} \Omega'(\zeta) &= pR \qquad (|\zeta| < 1) \\ &= \frac{pRmn}{\zeta^{n+1}} \quad (|\zeta| > 1) \end{aligned} \quad (5.78)$$

and from (5.77)

$$\omega'(\zeta) = \frac{pR}{\zeta^2} - pRmn\frac{d}{d\zeta}\left\{\frac{(1 + m\zeta^{n+1})}{(\zeta^{n+1} - mn)}\frac{1}{\zeta}\right\}$$

which agrees with the results (5.59), (5.60) obtained by the direct method of solution.

2. *Elliptical hole under tension*

As a second illustration which covers the case of known stresses at infinity, we consider an unstressed elliptical hole in an infinite plate under a known uniaxial tension T at infinity. If the mapping function is

$$z = m(\zeta) = R(\zeta + m\zeta^{-1}) \quad (0 \leqslant m < 1, |\zeta| \geqslant 1)$$

and the tension acts at an angle α to the real axis in the z-plane then

$$\Omega'(\zeta) = \frac{RT}{4} + O(\zeta^{-2}), \qquad \omega'(\zeta) = -\frac{RTe^{-2i\alpha}}{2} + O(\zeta^{-2}) \quad (5.79)$$

for large $|\zeta|$. From the examination of the singularities of $\Omega'(\zeta)$ we see it has a pole of order 2 at $\zeta = 0$. Then, since the hole is unstressed, on substituting in (5.74)

$$\Omega'(\zeta) = \frac{A_2}{\zeta^2} + \frac{A_1}{\zeta} + A_0.$$

From (5.79) $A_0 = RT/4$ and $A_1 = 0$. The remaining constant follows from the holomorphy condition (5.73) which requires

$$\omega'(\zeta) = \frac{1}{\zeta^2}\left(\frac{RT}{4} + \bar{A}_2\zeta^2\right) - \frac{d}{d\zeta}\left\{\frac{(1 + m\zeta^2)\zeta}{(\zeta^2 - m)}\left(\frac{RT}{4} + \frac{A_2}{\zeta^2}\right)\right\}$$

be holomorphic for $|\zeta| \geqslant 1$ and behave as (5.79) for large $|\zeta|$.
Hence

$$\bar{A}_2 - \frac{mRT}{4} = -\frac{RT}{2}e^{-2i\alpha}$$

so that

$$\Omega'(\zeta) = \frac{RT}{4} + \frac{RT}{4}(m - 2e^{2i\alpha})\frac{1}{\zeta^2} \quad (5.80)$$

agreeing with the result (5.54) obtained previously.

It is of interest to examine the stress field around this elliptical hole in some detail. Since $\tau_{\xi\xi} + \tau_{\eta\eta} = 4\,\mathrm{Re}\{\Omega'(\zeta)/m'(\zeta)\}$ and $\tau_{\xi\xi} = 0$ on the hole it is a simple matter to calculate the hoop stress $\tau_{\eta\eta}$ around the hole. We find

$$\tau_{\eta\eta} = \mathrm{Re}\left\{\frac{T(\zeta^2 + m - 2e^{2i\alpha})}{\zeta^2 - m}\right\}$$

on $\zeta = e^{i\eta}$ so that

$$\tau_{\eta\eta} = T\left(\frac{1 - m^2 + 2m\cos 2\alpha - 2\cos 2(\eta - \alpha)}{1 - 2m\cos 2\eta + m^2}\right)$$

at the point $z = R(e^{i\eta} + me^{-i\eta})$ on the elliptic boundary.

This expression may be differentiated to determine the positions of the maximum and minimum values of $\tau_{\eta\eta}$. In the particular case when the tension acts parallel to the minor axis of the hole ($\alpha = \frac{1}{2}\pi$) the maximum value of $\tau_{\eta\eta}$ is attained at the ends of the major axis and is

$$T\left(\frac{3+m}{1-m}\right) = T\left(1 + \frac{2a}{b}\right)$$

where a, b are the major and minor semi-axes of the hole. We note that if $b \to 0$ the stress concentration factor

$$\frac{\max \tau_{\eta\eta}}{T} = 1 + \frac{2a}{b}$$

tends to infinity, indicating the presence of high stresses near the ends of a thin elliptical slit or crack in a body.

In some problems it may be convenient to express the complex potentials in terms of the Cartesian coordinates z rather than the curvilinear coordinates $\zeta = \xi e^{i\eta}$. This is only possible if the inverse transformation $\zeta = m^{-1}(z)$ is of a simple form. In the present case

$$z = R(\zeta + m\zeta^{-1})$$

so that

$$\zeta = \left\{\frac{z + (z^2 - 4R^2m)^{1/2}}{2R}\right\}$$

where the sign of the square root was chosen to conform with $z \approx R\zeta$ for large $|\zeta|$. On substituting from (5.80), the complex potential $\Omega'(z)$ has the form

$$\Omega'(z) = \frac{\Omega'(\zeta)}{m'(\zeta)} = \frac{T}{4} - \frac{T(1 + e^{2i\alpha})}{4m}\left\{\frac{z}{(z^2 - 4R^2m)^{1/2}} - 1\right\} \qquad (5.81)$$

after some manipulation. The constants R, m in (5.81) are related to the major and minor semi-axes a, b by

$$m = (a - b)/(a + b), \qquad 4R^2m = a^2 - b^2.$$

In the particular case when $m = 1$ and $b = 0$ this potential and the associated $\omega'(z)$ should be compared with the solution to the corresponding crack problem which may be derived following the results of Section 3.10. In most cases it is algebraically more straight-forward to deal with the curvilinear coordinates $\zeta = \xi e^{i\eta}$.

3. *Epitrochoidal disc*

As an example involving a transformation of a curvilinear disc onto the region $|\zeta| \leqslant 1$ we consider the mapping (5.7)

$$z = R(\zeta + m\zeta^n) \quad (0 \leqslant m < 1/n)$$

in which the disc has as its boundary an epitrochoid. Let us suppose the displacement is specified on the boundary of the disc so that

$$u + \mathrm{i}v = D(t) \quad (t = \mathrm{e}^{\mathrm{i}\eta})$$

at the point defined by $z = R(\mathrm{e}^{\mathrm{i}\eta} + m\mathrm{e}^{\mathrm{i}n\eta})$ on the epitrochoid. On using the displacement continuation, which from (5.63) implies

$$\omega(\zeta) = \kappa\, \overline{\Omega(1/\bar{\zeta})} - \frac{\overline{m(1/\bar{\zeta})}}{m'(\zeta)}\, \Omega'(\zeta) \quad (\zeta \in S^+) \tag{5.82}$$

and substituting into the boundary conditions we find

$$\kappa(\Omega^+(t) - \Omega^-(t)) = D(t) \quad \text{on } |t| = 1.$$

Again this is a Cauchy problem for a circular boundary where the function $\Omega(\zeta)$ is holomorphic in $|\zeta| \leqslant 1$ but may have isolated poles in $|\zeta| > 1$. In fact, in this case, as $\Omega(\zeta)$ has the same poles as $m(\zeta)$, it has a pole of order n at infinity. Thus the Cauchy problem has the solution

$$\Omega(\zeta) = \frac{1}{2\pi \mathrm{i}\kappa} \int_\Gamma \frac{D(t)\,\mathrm{d}t}{t - \zeta} + A_0 + A_1\zeta + \cdots + A_n\zeta^n \tag{5.83}$$

where Γ is the contour $|t| = 1$ taken in an anticlockwise manner. The holomorphy condition in this case is that $\omega(\zeta)$ defined by (5.82) should be holomorphic in $|\zeta| \leqslant 1$.

As a simple example, let us suppose $D(t) = \varepsilon t$ where ε is real. Then since

$$\frac{1}{2\pi\mathrm{i}} \int_\Gamma \frac{t\,\mathrm{d}t}{t - \zeta} \begin{cases} = 0 & (|\zeta| > 1) \\ = \zeta & (|\zeta| < 1) \end{cases}$$

$\Omega(\zeta)$ is given by (5.83) and hence, from (5.82), $\omega(\zeta)$ has the form

$$\omega(\zeta) = \kappa\left(\bar{A}_0 + \frac{\bar{A}_1}{\zeta} + \cdots + \frac{\bar{A}_n}{\zeta^n}\right) \\ - \frac{1}{\zeta^n}(m + \zeta^{n-1})(1 - mn\zeta^{n-1})\left(\frac{\varepsilon}{\kappa} + A_1 + \cdots + nA_n\zeta^{n-1}\right).$$

This function is holomorphic at $\zeta = 0$ if

$$\kappa\bar{A}_1 = mnA_n + (1 - m^2 n)\left(\frac{\varepsilon}{\kappa} + A_1\right)$$

$$\kappa\bar{A}_n = m\left(\frac{\varepsilon}{\kappa} + A_1\right)$$

$$msA_s = \kappa A_{n-s+1} \quad (s = 2, 3, \ldots, n - 1).$$

When ε is real these equations have the simple solution

$$A_1 = -\frac{m^2 n\varepsilon}{\kappa(\kappa + m^2 n)}, \qquad A_n = \frac{m\varepsilon}{\kappa(\kappa + m^2 n)},$$

$$A_s = 0 \quad (s = 2, 3, \ldots, n - 1).$$

Thus

$$\Omega(\zeta)\begin{cases} = \dfrac{\varepsilon\zeta}{\kappa + m^2 n}\left(1 + \dfrac{m}{\kappa}\zeta^{n-1}\right) & (|\zeta| < 1) \\ = \dfrac{\varepsilon m\zeta}{\kappa(\kappa + m^2 n)}(-mn + \zeta^{n-1}) & (|\zeta| > 1) \end{cases}$$

from which the stress and displacement fields may be evaluated.

For a displacement boundary value problem for the region exterior to a boundary C the complex potentials contain logarithmic terms. On transforming to the region $|\zeta| \geqslant 1$ and making the displacement continuation (5.82) we find $\Omega(\zeta)$ satisfies the boundary conditions

$$\kappa\{\Omega^+(t) - \Omega^-(t)\} = D(t) \quad \text{on } |t| = 1.$$

As in the earlier chapters this boundary condition is most easily satisfied in differentiated form $\kappa\{\Omega'^+(t) - \Omega'^-(t)\} = D'(t)$ on $|t| = 1$ which effectively removes the multiple-valued logarithmic terms from the problem. The solution now proceeds as above and a more general example of this type is given in the next section.

5.6 The mixed boundary value problem

In this case we assume the complex displacement $D = u + iv$ is given on a set of arcs L of the boundary C of the region R_z occupied by the body, the remainder L' of C being unstressed. On transforming to the region S^+, namely $|\zeta| \geqslant 1$ or $|\zeta| \leqslant 1$ of the ζ-plane, the boundary conditions become

$$\begin{aligned} D = u + iv &= D(t) \quad (t \in L) \\ \tau_{\xi\xi} + i\tau_{\xi\eta} &= 0 \qquad\ \ (t \in L') \end{aligned} \tag{5.84}$$

on the sets of arcs L and L' of unit circle $|t| = 1$ which correspond to L and L' of C. On making the stress continuation (5.64) the boundary conditions reduce to

$$\kappa\,\Omega^+(t) + \Omega^-(t) = D(t) \quad (t \in L)$$

where $\Omega(\zeta)$ is holomorphic in the whole plane cut along L except at the set of poles described in Section 5.4 which lie in S^-. From (5.66) on putting $\zeta = \xi e^{i\eta}$ and differentiating with respect to η we find

$$\begin{aligned} 2\mu\, D'(\zeta)i\zeta = \kappa\,\Omega'(\zeta)i\zeta + \Omega'\left(\frac{1}{\bar{\zeta}}\right)\frac{i}{\bar{\zeta}} - \left\{m(\zeta) - m\left(\frac{1}{\bar{\zeta}}\right)\right\}\frac{d}{d\eta}\left\{\overline{\frac{\Omega'(\zeta)}{m'(\zeta)}}\right\} \\ - \left\{m'(\zeta)i\zeta - m'\left(\frac{1}{\bar{\zeta}}\right)\frac{i}{\bar{\zeta}}\right\}\overline{\frac{\Omega'(\zeta)}{m'(\zeta)}}. \end{aligned}$$

Hence on the boundary $|\zeta| = 1$

$$2\mu\, D'(t) = \kappa\, \Omega'^{+}(t) + \Omega'^{-}(t). \tag{5.85}$$

Thus $\Omega'(\zeta)$ is a single-valued function satisfying the conditions described above and the Hilbert problem (5.85) on the circle $t\bar{t} = 1$. As in Section 4.6 the solution to the problem is

$$\Omega'(\zeta) = \frac{X(\zeta)}{2\pi \mathrm{i}} \int_L \frac{2\mu\, D'(t)\, \mathrm{d}t}{X^{+}(t)(t - \zeta)} + \psi(\zeta)X(\zeta) \tag{5.86}$$

where $X(\zeta)$ is the basic Plemelj function for the arcs L of the unit circle and has been defined in (4.76) and (4.77) (see also Section 1.6). In this case, since the integral in (5.86) defines a function which is holomorphic in the whole plane cut along the arcs L and has $O(\zeta^{-n-1})$ as $|\zeta| \to \infty$, the function $\psi(\zeta)$ must have exactly the same singularities as $\Omega'(\zeta)$ at its isolated poles in S^{-}, must be such that $\Omega'(\zeta)$ has the correct behaviour at the origin and infinity, and be holomorphic elsewhere in the whole plane. Clearly once the behaviour of $\Omega'(\zeta)$ is known it is possible to write down the form of $\psi(\zeta)$ in terms of the principal parts of the poles. Again some of the constants in $\psi(\zeta)$ may be determined on substituting in the holomorphy condition of (5.73), the remainder come from additional physical assumptions which are, for example, that the resultant force acting over each arc of L is known, or that the relative depths of penetration of the punches which cause the displacement are known. Assumptions of this nature have been found necessary in the previous discussions of the mixed problem (Sections 3.7, 4.6).

As an illustration we consider an infinite plate containing an elliptical hole where a portion L of the hole is strongly reinforced so that it may only move as a rigid body (we shall for simplicity assume $D(t) =$ constant, corresponding to a rigid-body translation), the plate being deformed by a known uniaxial tension T at infinity acting at an angle α to the major axis of the ellipse.

We map the plate onto the region $|\zeta| \geqslant 1$ by the mapping

$$z = R(\zeta + m\zeta^{-1}) \quad (0 \leqslant m < 1).$$

From the discussion of Section 5.4 we see that $\Omega'(\zeta)$ has a pole of order 2 at $\zeta = 0$ and behaves at infinity according to (5.70) as

$$\Omega'(\zeta) = \frac{RT}{4} - \frac{(X + \mathrm{i}Y)}{2\pi(1 + \kappa)} \frac{1}{\zeta} + O\left(\frac{1}{\zeta^2}\right).$$

Also

$$\omega'(\zeta) = -\frac{RT}{2} \mathrm{e}^{-2\mathrm{i}\alpha} + \frac{\kappa(X - \mathrm{i}Y)}{2\pi(1 + \kappa)} \frac{1}{\zeta} + O\left(\frac{1}{\zeta^2}\right) \quad \text{for large } |\zeta| \tag{5.87}$$

where $X + iY$ is the resultant force over the hole, or equivalently the resultant force applied to the arc L, and we shall assume this force to be known. Let us suppose the part $|\eta| \leqslant \phi$ of the circle $|\zeta| = 1$ corresponds to the region L of the ellipse that is reinforced, so that the reinforcement is symmetrical about one end of the major axis (see Figure 5.4). We assume

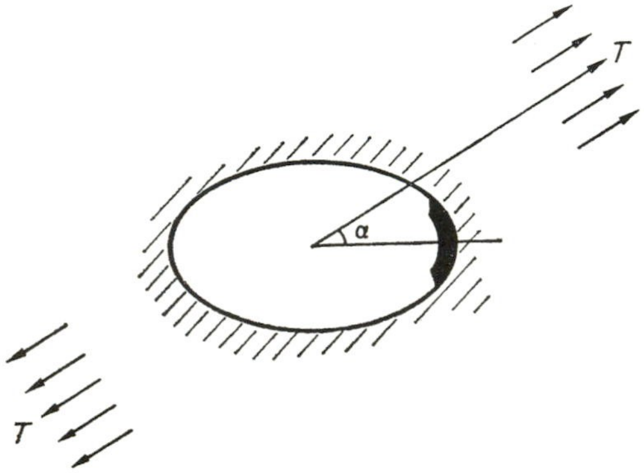

Figure 5.4

$D(t) =$ constant on $|\eta| \leqslant \phi$ so that $\Omega'(\zeta)$ satisfies the Hilbert problem

$$\kappa\, \Omega'^{+}(t) + \Omega'^{-}(t) = 0 \quad \text{for } t = e^{i\eta}, \quad |\eta| \leqslant \phi.$$

Hence from (5.86) $\Omega'(\zeta)$ has the form $\Omega'(\zeta) = \psi(\zeta)X(\zeta)$

where
$$X(\zeta) = (\zeta - e^{i\phi})^{-\gamma}(\zeta - e^{-i\phi})^{\gamma-1}$$

$$\gamma = \frac{1}{2} + i\beta = \frac{1}{2} + \frac{i}{2\pi}\log \kappa \tag{5.88}$$

$$\lim_{|\zeta| \to \infty} \zeta\, X(\zeta) = 1$$

and $\psi(\zeta)$ must be chosen to fit the known behaviour of $\Omega'(\zeta)$. Thus $\psi(\zeta)$ must have the form

$$\frac{B_2}{\zeta^2} + \frac{B_1}{\zeta} + A_0 + A_1\zeta. \tag{5.89}$$

For large $|\zeta|$, on using the expansion for $X(\zeta)$ derived in Section 4.6 or Section 1.7, we find

$$\Omega'(\zeta) = \left(A_1\zeta + A_0 + \frac{B_1}{\zeta} + \frac{B_2}{\zeta^2}\right)\left(\frac{1}{\zeta} + \frac{\gamma e^{i\phi} + (1 - \gamma)e^{-i\phi}}{\zeta^2} + \cdots\right)$$

so that on comparison with (5.87)

$$A_1 = \tfrac{1}{4}RT$$

$$A_0 + A_1\{\gamma e^{i\phi} + (1 - \gamma)e^{-i\phi}\} = -\frac{X + iY}{2\pi(1 + \kappa)}. \tag{5.90}$$

The remaining constants are most easily found on substituting into the holomorphy condition (5.73) giving

$$\omega'(\zeta) = \frac{1}{\zeta^2}\left(\bar{B}_2\zeta^2 + \bar{B}_1\zeta + \bar{A}_0 + \frac{\bar{A}_1}{\zeta}\right)\overline{X\left(\frac{1}{\bar{\zeta}}\right)}$$
$$- \frac{d}{d\zeta}\left[\frac{(1 + m\zeta^2)\zeta}{(\zeta^2 - m)}\left\{\frac{B_2}{\zeta^2} + \frac{B_1}{\zeta} + A_0 + A_1\zeta\right\}X(\zeta)\right].$$

This function must be holomorphic for $|\zeta| \geqslant 1$ and have the correct behaviour at infinity from (5.87). As this function is holomorphic for $|\zeta| \geqslant 1$ we need satisfy only the conditions at infinity by taking the expansion for large $|\zeta|$ which yields

$$\omega'(\zeta) = -e^{2\phi\beta}\left[\bar{B}_2 + \frac{\bar{B}_1 + \bar{B}_2\{\gamma e^{i\phi} + (1 - \gamma)e^{-i\phi}\}}{\zeta}\right] - mA_1 + O(\zeta^{-2}).$$

The first term in the expansion resulting from the form of $X(\zeta)$ for small $|\zeta|$ (see Section 1.7). On comparison with (5.87) we find

$$B_2 = -\frac{RT}{4}(m - 2e^{2i\alpha})e^{-2\phi\beta}$$
$$B_1 = -\frac{\kappa(X + iY)}{2\pi(1 + \kappa)}e^{-2\phi\beta} - B_2\{\gamma e^{-i\phi} + (1 - \gamma)e^{i\phi}\}. \tag{5.91}$$

Thus all constants have been found and $\Omega'(\zeta)$ may be evaluated. In the particular case when $m = 0$ and $X + iY = 0$ the solution reduces to that obtained in Section 4.6. It will also be noticed that the solution for the reinforced arc under a force $X + iY$ with zero stresses at infinity has the simple form

$$\Omega'(\zeta) = -\frac{(X + iY)}{2\pi(1 + \kappa)}\left(1 + \frac{\kappa e^{-2\phi\beta}}{\zeta}\right)(\zeta - e^{i\phi})^{-\gamma}(\zeta - e^{-i\phi})^{\gamma - 1} \tag{5.92}$$

the constants R, m which occur in the transformation are not present in this result but merely influence the value of ϕ. It should be remarked that as the reinforced arc was not allowed to rotate in the above calculations, this solution is only directly applicable in the symmetric case when $Y = 0$.

5.7 Frictionless boundary conditions

As mentioned in Section 4.8, many problems involving inclusions or inserts into an elastic body reduce to boundary value problems in which the normal displacement u_ξ, and the shear stress $\tau_{\xi\eta}$ are specified over the whole of the boundary contour $\xi = 1$ of the region R_z occupied by the body. If we transform the region R_z into the region S^+ of the ζ-plane, bounded by the contour $|\zeta| = 1$, the relations with the complex potentials reduce to (5.25) and (5.26) namely

$$\begin{aligned}\operatorname{Im}\left\{-\frac{\bar{\zeta}}{\zeta\, m'(\zeta)}\left[m(\zeta)\frac{\mathrm{d}}{\mathrm{d}\zeta}\overline{\left\{\frac{\Omega'(\zeta)}{m'(\zeta)}\right\}} + \overline{\omega'(\zeta)}\right]\right\} &= \tau_{\xi\eta}\\ \operatorname{Re}\left\{\overline{\zeta\, m'(\zeta)}\left[\kappa\, \Omega(\zeta) - \frac{m(\zeta)}{\overline{m'(\zeta)}}\overline{\Omega'(\zeta)} - \overline{\omega(\zeta)}\right]\right\} &= 2\mu|\zeta\, m'(\zeta)|u_\xi.\end{aligned} \tag{5.93}$$

If, on the boundary $t\bar{t} = 1$, $u_\xi = f(t)$, $\tau_{\xi\eta} = g(t)$, the boundary conditions take the form

$$\frac{1}{t^2}\overline{m'(t)}\,\{m(t)\,\overline{W'(t)} + \overline{\omega'(t)}\} - t^2 m'(t)\{\overline{m(t)}W'(t) + \omega'(t)\}$$
$$= -2\mathrm{i}|m'(t)|^2 g(t) \quad \text{on } t\bar{t} = 1 \tag{5.94}$$

$$\frac{\overline{m'(t)}}{t}[\kappa\, \Omega(t) - m(t)\overline{W(t)} - \overline{\omega(t)}] + t\, m'(t)[\kappa\, \overline{\Omega(t)} - \overline{m(t)}W(t) - \omega(t)]$$
$$= 4\mu|m'(t)|f(t) \quad \text{on } t\bar{t} = 1 \tag{5.95}$$

where $$W(\zeta) = \Omega'(\zeta)/m'(\zeta). \tag{5.96}$$

The functions $\Omega(\zeta)$, $\omega(\zeta)$ are holomorphic in S^+, and $m(\zeta)$ is a known rational function in the whole plane having poles in S^-. Our problem is to determine the functions $\Omega(\zeta)$, $\omega(\zeta)$ which satisfy the boundary conditions (5.94), (5.95). One approach is to use the direct method as in Section 4.8 and to integrate through each equation with

$$\frac{1}{2\pi\mathrm{i}}\int_\Gamma \frac{\mathrm{d}t}{t-\zeta} \quad \text{for } \zeta \in S^+.$$

The resulting integrals, though complicated, may be easily evaluated using the known properties of the functions and hence $\Omega(\zeta)$, $\omega(\zeta)$ may be found. This method is perfectly feasible but, if $m(\zeta)$ contains several poles in S^-, may be algebraically more complicated than the following method due to Muskhelishvili,[43] Section 128. Muskhelishvili's approach is to choose two representations for $\omega(\zeta)$ so that the boundary conditions (5.94) and (5.95) reduce to Cauchy boundary conditions.

We denote the inverse region of S^+ in $|\zeta| = 1$ by S^- and the circle $|\zeta| = 1$ by Γ oriented so that S^+ lies to the left. To define the continuation corresponding to $\tau_{\xi\eta}$ let us assume $\tau_{\xi\eta} = 0$ on $|\zeta| = 1$, then from (5.94) in the usual notation

$$t^2m'(t)\{\overline{m(t)}W'^+(t) + \omega'^+(t)\} = \frac{1}{t^2}\overline{m'(t)}\{m(t)\overline{W'^+(t)} + \overline{\omega'^+(t)}\}$$
$$= \frac{1}{t^2}\overline{m'\left(\frac{1}{\bar{t}}\right)}\left\{m(t)\overline{W'^-\left(\frac{1}{\bar{t}}\right)} + \overline{\omega'^-\left(\frac{1}{\bar{t}}\right)}\right\}$$

and hence the function

$$\theta(\zeta)\begin{cases} = \zeta^2 m'(\zeta)\left\{\overline{m\left(\frac{1}{\bar{\zeta}}\right)}W'(\zeta) + \omega'(\zeta)\right\} & (\zeta \in S^+) \\ = \frac{1}{\zeta^2}\overline{m'\left(\frac{1}{\bar{\zeta}}\right)}\left\{m(\zeta)\overline{W'\left(\frac{1}{\bar{\zeta}}\right)} + \overline{\omega'\left(\frac{1}{\bar{\zeta}}\right)}\right\} & (\zeta \in S^-) \end{cases} \tag{5.97}$$

is holomorphic in S^+ and S^- (except possibly at the origin and infinity and the poles of $m(\zeta)$, $\overline{m(1/\bar{\zeta})}$) and is continuous across $|\zeta| = 1$ when $\tau_{\xi\eta} = 0$. On inverting the second relation of (5.97) we obtain

$$\overline{\theta\left(\frac{1}{\bar{\zeta}}\right)} = \zeta^2 m'(\zeta)\left[\overline{m\left(\frac{1}{\bar{\zeta}}\right)}W'(\zeta) + \omega'(\zeta)\right] \quad \text{for } \zeta \in S^+$$

and hence $\theta(\zeta)$ must be such that

$$\overline{\theta\left(\frac{1}{\bar{\zeta}}\right)} = \theta(\zeta) \quad \text{for } \zeta \in S^+. \tag{5.98}$$

Again we note the first relation of (5.97) could be regarded as a convenient representation for $\omega'(\zeta)$ when $\theta(\zeta)$ satisfies (5.98). On substituting from (5.97) into (5.93) we find

$$-2i\tau_{\xi\eta} = \frac{1}{|\zeta\, m'(\zeta)|^2}\left\{\overline{\theta\left(\frac{1}{\bar{\zeta}}\right)} - \theta(\zeta)\right\} + \frac{\bar{\zeta}}{\zeta\, m'(\zeta)}\left\{m(\zeta) - m\left(\frac{1}{\bar{\zeta}}\right)\right\}\overline{W'(\zeta)}$$
$$- \frac{\zeta}{\overline{\zeta\, m'(\zeta)}}\left\{\overline{m(\zeta)} - \overline{m\left(\frac{1}{\bar{\zeta}}\right)}\right\}W'(\zeta) \tag{5.99}$$

and hence the boundary condition (5.94) becomes

$$\theta^+(t) - \theta^-(t) = 2i|m'(t)|^2 g(t) \quad \text{on } t\bar{t} = 1. \tag{5.100}$$

This is a simple Cauchy problem for the function $\theta(\zeta)$, where $\theta(\zeta)$ has its behaviour in the whole plane including the origin and infinity and at each pole defined by (5.97) and as a consequence must satisfy (5.98). Thus $\theta(\zeta)$ satisfies the Cauchy boundary conditions (5.100) on the circle and hence from Section 1.6 has the general representation

$$\theta(\zeta) = \frac{1}{\pi}\int_\Gamma \frac{|m'(t)|^2 g(t)\,dt}{t - \zeta} + \psi_1(\zeta) \tag{5.101}$$

where $\psi_1(\zeta)$ is an arbitrary function. Since the Cauchy integral in (5.101) is holomorphic in S^+ and S^-, $\psi_1(\zeta)$ has exactly the same distribution of poles as $\theta(\zeta)$ and is holomorphic elsewhere in the plane, having a known behaviour at the origin and infinity. Hence a knowledge of the positions and orders of the poles of $\theta(\zeta)$ and its properties at the origin and infinity enables us to write down the form of the complete representation for the function $\psi_1(\zeta)$ in terms of the principal parts at these poles.

Similarly it is possible to define the continuation corresponding to u_r by putting $u_r = 0$ in (5.96). In this case the function $\phi(\zeta)$ defined by

$$\phi(\zeta)\begin{cases} = \dfrac{\overline{m'(1/\bar{\zeta})}}{\zeta}\,\kappa\,\Omega(\zeta) - \zeta\, m'(\zeta)\left\{\overline{m\left(\dfrac{1}{\bar{\zeta}}\right)}W(\zeta) + \omega(\zeta)\right\} & (\zeta \in S^+) \\ = -\zeta\, m'(\zeta)\kappa\,\overline{\Omega\left(\dfrac{1}{\bar{\zeta}}\right)} + \dfrac{1}{\zeta}\overline{m'\left(\dfrac{1}{\bar{\zeta}}\right)}\left\{m(\zeta)\overline{W\left(\dfrac{1}{\bar{\zeta}}\right)} + \overline{\omega\left(\dfrac{1}{\bar{\zeta}}\right)}\right\} & (\zeta \in S^-) \end{cases} \tag{5.102}$$

is obtained. From these relations we see that $\phi(\zeta)$ is holomorphic in S^+ and S^- except possibly at the origin and infinity and the points corresponding to the poles of $m(\zeta)$, $\overline{m(1/\bar{\zeta})}$ and is such that

$$\overline{\phi\left(\frac{1}{\bar{\zeta}}\right)} = -\,\phi(\zeta) \quad \text{for } \zeta \in S^+. \tag{5.103}$$

On using the relations (5.102) to define a representation for $\omega(\zeta)$ and substituting in (5.93) we find

$$\begin{aligned} 4\mu|\zeta\, m'(\zeta)|u_\xi = \operatorname{Re}\Bigg[\left\{\overline{\zeta\, m'(\zeta)} - \frac{\overline{m'(1/\bar{\zeta})}}{\zeta}\right\}\kappa\,\Omega(\zeta) \\ -\overline{\zeta\, m'(\zeta)}\left\{m(\zeta) - \overline{m\left(\frac{1}{\bar{\zeta}}\right)}\right\}\overline{W(\zeta)}\Bigg] + \phi(\zeta) - \overline{\phi\left(\frac{1}{\bar{\zeta}}\right)}. \end{aligned} \tag{5.104}$$

Hence the boundary condition (5.95) on $|\zeta| = 1$ becomes

$$\phi^+(t) - \phi^-(t) = 4\mu|m'(t)|f(t). \tag{5.105}$$

Again this is a simple Cauchy problem with the solution

$$\phi(\zeta) = \frac{2\mu}{\pi \mathrm{i}}\int_\Gamma \frac{|m'(t)|f(t)\,\mathrm{d}t}{t - \zeta} + \psi_2(\zeta) \tag{5.106}$$

where $\psi_2(\zeta)$ is holomorphic in the whole plane, except possibly at the origin and infinity and the poles of $\phi(\zeta)$, where it is chosen to fit the behaviour of $\phi(\zeta)$. We note in addition that $\psi_1(\zeta)$, $\psi_2(\zeta)$ in (5.101) and (5.106) must be chosen so that equations (5.98) and (5.103) are satisfied.

To complete the solution it is necessary to determine $\Omega(\zeta)$, $\omega(\zeta)$ in terms of $\theta(\zeta)$, $\phi(\zeta)$ from (5.97) and (5.102). On eliminating $\omega'(\zeta)$ from these equations by differentiating (5.102) and remembering $W(\zeta) = \Omega'(\zeta)/m'(\zeta)$ we find $\Omega(\zeta)$ satisfies the linear differential equation

$$\frac{\mathrm{d}}{\mathrm{d}\zeta}\{\kappa Q(\zeta)\,\Omega(\zeta)\} + Q(\zeta)\,\Omega'(\zeta) = \frac{\theta(\zeta)}{\zeta^2 m'(\zeta)} + \frac{\mathrm{d}}{\mathrm{d}\zeta}\left\{\frac{\phi(\zeta)}{\zeta\, m'(\zeta)}\right\}$$

or

$$\Omega'(\zeta) + \frac{\kappa Q'(\zeta)}{(\kappa + 1)Q(\zeta)}\,\Omega(\zeta) = F(\zeta) \tag{5.107}$$

where

$$Q(\zeta) = \frac{\overline{m'(1/\bar{\zeta})}}{\zeta^2 m'(\zeta)}$$

and

$$F(\zeta) = \frac{1}{(\kappa + 1)\overline{m'(1/\bar{\zeta})}}\left[\theta(\zeta) + \zeta^2 m'(\zeta)\frac{\mathrm{d}}{\mathrm{d}\zeta}\left\{\frac{\phi(\zeta)}{\zeta\, m'(\zeta)}\right\}\right]. \tag{5.108}$$

The differential equation (5.107) should be compared with the corresponding equation (4.99) for a circular region. Again it proves possible to integrate (5.107) to give

$$\Omega(\zeta) = \{Q(\zeta)\}^{-\nu}\{\alpha + \int F(\zeta)\{Q(\zeta)\}^{\nu}\,\mathrm{d}\zeta\} \tag{5.109}$$

where α is a constant and $\nu = \kappa/(\kappa + 1)$ takes values in the range $\frac{1}{2} < \nu < 1$. Having found $\Omega(\zeta)$ we may determine $\omega(\zeta)$ from (5.102). The undetermined constants in $\psi_1(\zeta)$, $\psi_2(\zeta)$ must be chosen so that $\Omega(\zeta)$, $\omega(\zeta)$ have the correct properties in S^+. For instance, if S^+ is the region $|\zeta| \leqslant 1$ the complex potentials are holomorphic and single valued in $|\zeta| \leqslant 1$, or if S^+ is the region $|\zeta| \geqslant 1$, $\Omega'(\zeta)$, $\omega'(\zeta)$ should be holomorphic and single valued in $|\zeta| \geqslant 1$ and have the correct behaviour (5.18) at infinity. These conditions are sufficient to determine the potentials completely.

1. *Circular disc*

As a simple illustration, suppose we insert a circular elastic disc of radius a into a hole of radius $a - \varepsilon$ in a rigid plate and the materials make a frictionless contact. Then the transformation is $z = a\zeta$ and the boundary conditions are

$$u_\xi = -\varepsilon, \qquad \tau_{\xi\eta} = 0. \tag{5.110}$$

From the definition (5.97) since $\Omega(\zeta)$, $\omega(\zeta)$ are holomorphic in $|\zeta| \leqslant 1$ we see that $\theta(\zeta)$ is holomorphic in both S^+ and S^- and has zeros at the origin and infinity. Thus from (5.101) and the boundary conditions

$$\theta(\zeta) = \psi_1(\zeta) = 0.$$

Applying the same considerations to $\phi(\zeta)$ from (5.102), we find $\phi(\zeta)$ has a simple pole at the origin and infinity. Hence from (5.106) and (5.110)

$$\phi(\zeta) = -\frac{2\mu}{\pi \mathrm{i}}\int \frac{a\varepsilon \, \mathrm{d}t}{t-\zeta} + \frac{B_1}{\zeta} + A_0 + A_1\zeta$$

$$\phi(\zeta)\begin{cases} = -4\mu a\varepsilon + \dfrac{B_1}{\zeta} + A_0 + A_1\zeta & (|\zeta| < 1) \\ = \dfrac{B_1}{\zeta} + A_0 + A_1\zeta & (|\zeta| > 1). \end{cases}$$

Substituting these values in (5.103) gives

$$A_0 + \bar{A}_0 = 4\mu a\varepsilon$$
$$B_1 = -\bar{A}_1.$$

Now from (5.108) as

$$F(\zeta) = \frac{1}{(\kappa+1)a}\left(4\mu a\varepsilon - A_0 + \frac{2\bar{A}_1}{\zeta^2}\right)$$

we find

$$\Omega(\zeta) = \alpha\zeta^{2\nu} + \frac{(4\mu a\varepsilon - A_0)\zeta}{a(1-\kappa)} - \frac{\bar{A}_1}{a\kappa}$$

and hence for a single-valued function we must conclude $\alpha = 0$. We also see that $\omega(\zeta) = -A_1/a$ from (5.102). As the undetermined constants A_1 and $\mathrm{Im}(A_0)$ merely contribute to a rigid-body rotation they can be taken to be zero and the solution is simply

$$\Omega(\zeta) = \frac{2\mu\varepsilon\zeta}{(1-\kappa)}, \qquad \omega(\zeta) = 0 \tag{5.111}$$

agreeing with the result obtained in Section 4.8.

2. *Elliptical inclusion*

As a second illustration we consider an infinite medium containing a smooth rigid elliptical inclusion and suppose the medium is deformed by a uniaxial tension T at infinity which acts parallel to the minor axis of the inclusion. The transformation

$$z = R\left(\zeta + \frac{m}{\zeta}\right) \quad (0 \leqslant m < 1)$$

takes the medium onto the region $|\zeta| \geqslant 1$. The boundary conditions on the elliptical boundary are

$$u_\xi = \tau_{\xi\eta} = 0$$

since from symmetry the inclusion does not rotate. Also, the complex potentials $\Omega(\zeta)$, $\omega(\zeta)$ are holomorphic and single valued in $|\zeta| \geqslant 1$ having the forms

$$\Omega(\zeta) = \tfrac{1}{4}RT\zeta + O(1), \qquad \omega(\zeta) = \tfrac{1}{2}RT\zeta + O(1) \tag{5.112}$$

for large $|\zeta|$ from (5.18) as the resultant force $X + \mathrm{i}Y$ over the inclusion is zero. From (5.97) it will be seen that $\theta(\zeta)$ has poles of order 2 at infinity and the origin and hence from (5.101) has the form

$$\theta(\zeta) = \frac{R^2T}{2\zeta^2} + \frac{\alpha_{-1}}{\zeta} + \alpha_0 + \alpha_1\zeta + \frac{R^2T}{2}\zeta^2.$$

In fact a close inspection of (5.97) and (5.112) shows that $\alpha_{-1} = \alpha_1 = 0$. Further as $\theta(\zeta)$ must satisfy (5.98) the constants are such that

$$\alpha_0 = \bar{\alpha}_0 = A \quad \text{(real)}$$

so that

$$\theta(\zeta) = \frac{R^2T}{2\zeta^2} + A + \frac{R^2T}{2}\zeta^2. \tag{5.113}$$

Similarly, from the definition of $\phi(\zeta)$ from (5.102), $\phi(\zeta)$ also has poles of second order at the origin and infinity and thus from (5.106) and (5.103) has the form

$$\phi(\zeta) = -\frac{\bar{\beta}_2}{\zeta^2} - \frac{\bar{\beta}_1}{\zeta} + \mathrm{i}B + \beta_1\zeta + \beta_2\zeta^2 \tag{5.114}$$

where B is real and

$$\beta_2 = -\tfrac{1}{4}R^2T\{2 + (\kappa + 1)m\}. \tag{5.115}$$

A further very useful (and necessary) simplification can be introduced into (5.114) by noting that the addition of a complex constant β to $\Omega(\zeta)$ and $\kappa\bar{\beta}$ to $\omega(\zeta)$ does not affect the displacement field (in fact this is the only arbitrariness allowed in $\Omega(\zeta)$, $\omega(\zeta)$). On making these additions $\theta(\zeta)$ from (5.97) remains unchanged but $\phi(\zeta)$ from (5.102) is arbitrary to within a term of the form

$$R\kappa\{(\beta + m\bar{\beta})\zeta^{-1} - (\bar{\beta} + m\beta)\zeta\}.$$

As $0 \leqslant m < 1$ the arbitrary constant β may be chosen to eliminate the terms in β_1 from (5.114). Hence $\phi(\zeta)$ has the form

$$\phi(\zeta) = -\bar{\beta}_2\zeta^{-2} + \mathrm{i}B + \beta_2\zeta^2. \tag{5.116}$$

The only remaining unknown constants are A and B. On substituting in (5.109) we find $\Omega(\zeta)$ may be written in the form

$$\Omega(\zeta) = \alpha\left(\frac{\zeta^2 - m}{1 - m\zeta^2}\right)^\nu + \left(\frac{\zeta^2 - m}{1 - m\zeta^2}\right)^\nu \int_{-m^{-\frac{1}{2}}}^{\zeta} F(\zeta)\left(\frac{1 - m\zeta^2}{\zeta^2 - m}\right)^\nu \mathrm{d}\zeta \quad (5.117)$$

where the lower limit in the integral has been chosen for convenience and $F(\zeta)$ is given by

$$F(\zeta) = \frac{1}{R(\kappa + 1)(1 - m\zeta^2)}\left[\theta(\zeta) + (\zeta^2 - m)\frac{\mathrm{d}}{\mathrm{d}\zeta}\left\{\frac{\zeta\,\phi(\zeta)}{\zeta^2 - m}\right\}\right]. \quad (5.118)$$

The remaining constants must be chosen to ensure $\Omega(\zeta)$ is holomorphic and single valued in $|\zeta| \geqslant 1$.

On examining (5.117) it will be seen that the points $\zeta = \pm m^{-1/2}$ may be branch points of $\Omega(\zeta)$ in $|\zeta| \geqslant 1$, however the integral is convergent ($\frac{1}{2} < \nu < 1$) and the second term remains finite as $\zeta \to -m^{-1/2}$. Hence for $\Omega(\zeta)$ to remain finite we must conclude $\alpha = 0$. Similarly $\Omega(\zeta)$ is finite at $\zeta = m^{-1/2}$ if

$$\int_{-m^{-\frac{1}{2}}}^{m^{-\frac{1}{2}}} F(\zeta)\left(\frac{1 - m\zeta^2}{\zeta^2 - m}\right)^\nu \mathrm{d}\zeta = 0 \quad (5.119)$$

and it may now be confirmed that $\Omega(\zeta)$ is holomorphic in $|\zeta| \geqslant 1$.

Without going into the details it will be seen from (5.113) and (5.116), that (5.119) is a linear equation of the form

$$AJ_1 - \mathrm{i}BJ_2 = J_3$$

where

$$J_1 = \int_{-m^{-\frac{1}{2}}}^{m^{-\frac{1}{2}}} \frac{\mathrm{d}\zeta}{(1 - m\zeta^2)^{1-\nu}(\zeta^2 - m)^\nu}$$

$$J_2 = \int_{-m^{-\frac{1}{2}}}^{m^{-\frac{1}{2}}} \frac{(\zeta^2 + m)\,\mathrm{d}\zeta}{(1 - m\zeta^2)^{1-\nu}(\zeta^2 - m)^{\nu+1}}$$

$$J_3 = \int_{-m^{-\frac{1}{2}}}^{m^{-\frac{1}{2}}} \left[\left(\frac{\zeta^2 + m}{\zeta^2 - m}\right)\psi(\zeta) - \zeta\,\psi'(\zeta) - \frac{R^2T}{2}\left(\zeta^2 + \frac{1}{\zeta^2}\right)\right] \frac{\mathrm{d}\zeta}{(1 - m\zeta^2)^{1-\nu}(\zeta^2 - m)^\nu}$$

and

$$\psi(\zeta) = \beta_2\zeta^2 - \bar{\beta}_2\zeta^{-2}$$

from which A and B may be found. Finally $\omega(\zeta)$ may be determined from (5.102) and has the correct properties in $|\zeta| \geqslant 1$. Thus a complete solution to the problem has been obtained.

It is clear that the above solution may be generalized to the unsymmetric case where a non-zero force is applied to the inclusion with arbitrary stresses at infinity. As in Section 4.8 the problems of incomplete contact of a smooth rigid inclusion and an elastic medium are not reducible to a Hilbert problem and have been solved using numerical techniques. References have been given in Section 4.8.

5.8 Concluding remarks

As it has not proved possible for reasons of space to consider more general mapping techniques we conclude by commenting on some of the main approaches to this extensive field.

There are two basic limitations of conformal mapping techniques. Firstly a multiply connected region can only be conformally mapped onto one of like connectivity. Consequently the boundary value problems for multiply connected regions may only be solved (by mapping techniques) if solutions are known to the boundary value problems for representative multiply connected regions, for example, a circle, an annular region, a circle containing two holes, etc. Although solutions may be found for such representative regions it will be appreciated that the analytical difficulties increase with increasing connectivity. However, approximate and exact solutions have been obtained for some problems for bodies with infinite connectivities, such as an infinite plate containing a doubly periodic array of circular holes or an infinite plate containing an infinite row of collinear line cracks, as the geometry for these bodies is particularly suited to certain analytical techniques.

Secondly the problem of determining the mapping function to transform the given region into a representative one of like connectivity may be a major problem in its own right. Whilst extensive tables of mapping functions exist,[30] in a particular problem it may be necessary to make do with an approximate mapping function determined by a numerical technique.†

The use of approximate mapping functions has already been mentioned in Section 5.1 with regard to mapping the infinite plane containing a polygonal hole onto the exterior of the unit circle. In this case the exact mapping function is known from the Schwarz-Christoffel transformation but has branch points on the boundary of the unit circle. Such branch points may cause severe algebraic difficulties and it is common practice to approximate the mapping function by truncating the Laurent series of the

† Very recently the various methods of generating mapping functions have been compared by Kolchanov and Shvetsov.[32]

transformation. As indicated in Sections 5.3 and 5.5, if only a few terms are retained in the Laurent series the transformed problem may be solved by hand calculations. Several examples of this nature have been given by Savin.[50] If many more terms are retained with a consequent increase in accuracy of the solution, the resulting linear equations must be solved on a computer. The current literature contains many examples of this technique as applied to different geometries for which the exact mapping functions are known, see, for example, Bowie.[2]

This chapter has been devoted exclusively to the case of elastic regions which may be mapped by means of rational functions onto the interior (or exterior) of the unit circle. As an alternative approach we could consider bodies which map conformally onto a half-plane or an infinite strip. The solution to the corresponding problems for a half-plane follows on extending Chapter 3 in the manner described in Section 5.4. The corresponding problems for an infinite strip are less easily dealt with and recourse has to be made to either integral transform techniques or eigenfunction expansions.[4, 37, 65]

A method which is entirely equivalent to the conformal mapping technique is to choose a system of orthogonal curvilinear coordinates so that the boundaries of the elastic body correspond to coordinate lines. This method has been employed by Neuber[46] using elliptical and other coordinate systems to investigate the stresses in bodies containing hyperbolic notches. Several authors have used bipolar coordinates to solve a number of problems concerning regions having two circular boundaries. Green and Zerna[21] and Milne-Thomson[41] give illustrations of this method with appropriate references.

Examples on Chapter 5

1. Use the series method to solve the problem of an unstressed elliptical hole in an infinite plate under a uniform tension T at infinity acting at an angle α to the major axis of the hole. If the transformation is written in the form $z = R(\zeta + m/\zeta)$ confirm that

$$\Omega(\zeta) = \frac{RT}{4}\zeta + \frac{RT}{4}(2e^{2i\alpha} - m)\frac{1}{\zeta}$$

and that $\omega(\zeta)$ is given by (5.54).

2. Use the series method to confirm that the complex potentials

$$\Omega(\zeta) = \frac{2\mu\varepsilon Rmi}{\kappa\zeta}, \qquad \omega(\zeta) = \frac{2\mu\varepsilon Ri}{\zeta}\left\{1 + \frac{m}{\kappa}\left(\frac{1 + m\zeta^2}{\zeta^2 - m}\right)\right\}$$

correspond to an infinite plate bonded to a rigid elliptical inclusion which has been rotated through a small angle ε, the stress and rotation at infinity being zero. How is this solution changed if a uniform tension T acting at an angle α to the major axis of the inclusion is applied at infinity? Assuming the inclusion is free to rotate show that the tension T will cause it to rotate through the angle

$$\varepsilon = \frac{mT(\kappa + 1)\sin 2\alpha}{4\mu(\kappa + m^2)}.$$

3. Use the series method to show that the complex potentials for an epitrochoidal disc deformed under a uniform pressure p on its boundary are

$$\Omega(\zeta) = -\tfrac{1}{2}p\, m(\zeta), \qquad \omega(\zeta) = 0$$

where the disc is defined by the transformation

$$z = m(\zeta) = R(\zeta + m\zeta^n) \quad (|\zeta| \leqslant 1).$$

4. Show that the complex potentials

$$\Omega(\zeta) = -\tfrac{1}{2}p\, m(\zeta), \quad \omega(\zeta) = 0$$

give the solution corresponding to a uniform pressure p acting on the boundary of the general region defined by the mapping $z = m(\zeta)$ $(|\zeta| \leqslant 1)$, where $m(\zeta)$ is holomorphic and $m'(\zeta)$ non-zero in $|\zeta| \leqslant 1$.

5. Use the series method to show that the potentials corresponding to a hypotrochoidal hole in an infinite solid $z = R(\zeta + m/\zeta^n)$, $|\zeta| \geqslant 1$, inflated by a uniform pressure p with zero stress and rotation at infinity are

$$\Omega(\zeta) = -\frac{pRm}{\zeta^n}, \qquad \omega(\zeta) = -\frac{pR(1 + m^2 n)\zeta^n}{\zeta^{n+1} - mn}.$$

6. An infinite plate containing an elliptical hole is mapped onto $|\zeta| > 1$ by $z = R(\zeta + m/\zeta)$. If a uniform pressure p is applied to those parts of the hole corresponding to $\zeta = e^{i\theta}$ $(|\theta| < \theta_0,\ |\theta - \pi| < \theta_0)$ and the remainder of the hole is unstressed, show that

$$\Omega(\zeta) = \frac{p}{2\pi i}\left\{R\left(\zeta + \frac{m}{\zeta}\right)\log\left(\frac{\zeta^2 - \zeta_0^2}{\zeta^2 - \bar{\zeta}_0^2}\right) + z_0 \log\left(\frac{\zeta + \zeta_0}{\zeta - \zeta_0}\right) + \bar{z}_0 \log\left(\frac{\zeta - \bar{\zeta}_0}{\zeta + \bar{\zeta}_0}\right) - \frac{4iRm}{\zeta}\,\theta_0\right\}$$

where $z_0 = R(\zeta_0 + m/\zeta_0)$, $\zeta_0 = e^{i\theta_0}$. Hence show that the potential due to equal and opposite point forces X applied at the ends of the major axis of the ellipse is

$$\Omega(\zeta) = -\frac{X}{2\pi}\log\left(\frac{\zeta+1}{\zeta-1}\right).$$

What is the corresponding result for point forces at the ends of the minor axis?

7. An elastic disc with a reinforced boundary ($D = 0$) is deformed under the action of a finite number of point forces and moments acting at internal points. If the disc is mapped conformally onto the circle $|\zeta| < 1$ by $z = m(\zeta)$ then the point forces and moments may be represented by potentials $\Omega_0(\zeta)$, $\omega_0(\zeta)$ of the form of (3.6) in which the substitution $z = m(\zeta)$ has been made. On putting

$$\Omega(\zeta) = \Omega_0(\zeta) + \Omega_1(\zeta), \qquad \omega(\zeta) = \omega_0(\zeta) + \omega_1(\zeta)$$

show that the boundary conditions imply

$$-\kappa\,\Omega_1(\zeta) + \frac{m(\zeta)}{m'(1/\zeta)}\overline{\Omega_0'(1/\bar{\zeta})} + \overline{\omega_0(1/\bar{\zeta})} = \phi(\zeta) \qquad (|\zeta| \leqslant 1)$$

$$-\kappa\,\Omega_0(\zeta) + \frac{m(\zeta)}{m'(1/\zeta)}\overline{\Omega_1'(1/\bar{\zeta})} + \overline{\omega_1(1/\bar{\zeta})} = -\phi(\zeta) \quad (|\zeta| \geqslant 1)$$

where $\phi(\zeta)$ has poles at certain points but is holomorphic elsewhere in the whole plane. Indicate how $\phi(\zeta)$ may be determined and in particular confirm that when $m(\zeta) = R(\zeta + m\zeta^2)$ $(0 < m < \frac{1}{2})$, and $\Omega_0(\zeta)$, $\omega_0(\zeta)$ have $O(1/\zeta)$ as $|\zeta| \to \infty$

$$\phi(\zeta) = \alpha_1(\zeta + 2m)^{-1} + \beta_1\zeta + \beta_2\zeta^2$$

where $\alpha_1 = 4m^2(1-2m^2)\overline{\Omega_0'(-1/2m)}$. Find $\Omega(\zeta)$ and $\omega(\zeta)$ when there is a point moment M at the origin with $\Omega_0(\zeta) = 0$, $\omega_0(\zeta) = \mathrm{i}M/(2\pi R\zeta)$.

8. An infinite plate contains an elliptical hole, with major and minor axes a and b. If the plate is under an all-round tension and the hole is unstressed, determine the hoop stress around the hole and show that the stress concentration factors at the ends of the major and minor axes are $2a/b$ and $2b/a$ respectively. What are the stress concentration factors at these points when the plate is under an all-round tension and the hole is reinforced so that $D = 0$.

9. An infinite plate containing an equilateral triangular hole may be approximately mapped onto the region $|\zeta| \geqslant 1$ by

$$z = R\left\{\zeta + \frac{a_1}{\zeta^2} + \frac{a_2}{\zeta^5} + O\left(\frac{1}{\zeta^8}\right)\right\}$$

where $a_1 = \frac{1}{3}$, $a_2 = \frac{1}{45}$. If the hole is unstressed and a tension T acts at infinity at an angle α to the real axis, evaluate $\Omega(\zeta)$ for the three-term approximation given above. Hence determine the hoop stress at the point corresponding to $\zeta = 1$. As α varies show that this stress has a maximum of $3T$ for the one-term approximation ($a_1 = a_2 = 0$), $11T$ for the two-term approximation ($a_2 = 0$), and $17.9T$ for three-term approximation.

10. Use the continuation method to solve problems 2, 4, 8, 9.

11. An infinite plate with zero stress and rotation at infinity contains an elliptical hole. An arc of the hole defined by $z = R(\zeta + m/\zeta)$, $\zeta = e^{i\theta}$, $|\theta| < \phi$ is strongly reinforced so that it moves as a rigid body. If the rest of the hole is unstressed and the reinforced arc is acted on by a force with resultant $X + iY$ and is also rotated through a small angle ε show that

$$\Omega'(\zeta) = -\frac{X + iY}{2\pi(1+\kappa)}\left(1 + \frac{\kappa e^{-2\phi\beta}}{\zeta}\right)X(\zeta) + \frac{2iR\mu\varepsilon\kappa}{1+\kappa}\left(1 - \frac{m}{\zeta^2} + \left[-\zeta + \gamma e^{i\phi} + (1-\gamma)e^{-i\phi} + \frac{me^{-2\phi\beta}}{\zeta}\left\{\gamma e^{-i\phi} + (1-\gamma)e^{i\phi} - \frac{1}{\zeta}\right\}\right]X(\zeta)\right)$$

where

$$X(\zeta) = (\zeta - e^{i\phi})^{-\frac{1}{2}-i\beta}(\zeta - e^{-i\phi})^{-\frac{1}{2}+i\beta}, \qquad \beta = \frac{1}{2\pi}\log\kappa.$$

12. If the reinforced arc in example 11 is acted on by a symmetric force with resultant X show that the stress at the point $z = R(e^{i\eta} + me^{-i\eta})$ on the arc is given by

$$\tau_{\xi\xi} + i\tau_{\xi\eta} = -\frac{X}{2\pi R}\left(\frac{1 + \kappa e^{-2\phi\beta}e^{-i\eta}}{1 - me^{-2i\eta}}\right)X^{+}(e^{i\eta})$$

where

$$X^{+}(e^{i\eta}) = (R_1R_2)^{-1/2}(R_2/R_1)^{i\beta}e^{\beta(\phi-\pi)}e^{-i\eta/2}$$

and $R_1 = |e^{i\eta} - e^{i\phi}|$, $R_2 = |e^{i\eta} - e^{-i\phi}|$. Hence determine an upper bound to the length of arc over which the stress oscillates near the stress singularities.

BIBLIOGRAPHY

1. Barenblatt, G. I., *Advances in Applied Mechanics*, **7** (1962), 55.
2. Bowie, O. L., *J. appl. Mech.* **31** (1964), 208 and 726.
3. Buchwald, V. T., *Journal of the Australian Math. Soc.* **3** (1963), 93.
4. Buchwald, V. T., *Proc. Roy. Soc. A*, **277** (1964), 385.
5. Buchwald, V. T. and Davies, G. A. O., *Qt. Jl. Mech. appl. Math.* **17** (1964), 1.
6. Carathéodory, C., *Conformal Representation*, Cambridge, Cambridge University Press, 1932.
7. Carrier, G. F., Krook, M. and Pearson, C. E., *Functions of a Complex Variable*, New York, McGraw-Hill Book Co., 1966.
8. Churchill, R. V., *Introduction to Complex Variables*, New York, McGraw-Hill Book Co., 1960.
9. Cooke, J. C. and Tranter, C. J., *Qt. Jl. Mech. appl. Math.* **12** (1959), 379.
10. England, A. H., *J. appl. Mech.* **32** (1965), 400.
11. England, A. H., *J. appl. Mech.* **33** (1966), 637.
12. England, A. H. and Green, A. E., *Proc. Camb. phil. Soc.* **59** (1963), 489.
13. Epstein, B., *Partial Differential Equations*, New York, McGraw-Hill Book Co., 1962.
14. Erdélyi, A. (editor), *Tables of Integral Transforms*, Vols I and II, New York, McGraw-Hill Book Co., 1954.
15. Erdogan, F., *J. appl. Mech.* **32** (1965), 403.
16. Filon, L. N. G., *Phil. Trans. A*, **201** (1903), 63.
17. Galin, L. A., *Contact Problems in the Theory of Elasticity*, translation edited by I. N. Sneddon, Raleigh, North Carolina State College, 1961.
18. Goree, J. G., *J. appl. Mech.* **32** (1965), 437.
19. Green, A. E., *Proc. Roy. Soc. A*, **180** (1942), 173.
20. Green, A. E. and Adkins, J. E., *Large Elastic Deformations*, London, Oxford University Press, 1960.
21. Green, A. E. and Zerna, W., *Theoretical Elasticity*, London, Oxford University Press, 1954.
22. Griffith, A. A., *Phil. Trans. A*, **221** (1921), 163.
23. Gurtin, M. E., *Archs ration. Mech. Analysis*, **9** (1962), 225.
24. Gurtin, M. E., *Archs ration. Mech. Analysis*, **13** (1963), 321.
25. Irwin, G. R., *Fracture Mechanics*, First Symposium on Naval Structural Mechanics, New York, Pergamon Press, 1958.
26. Kantorowich, L. V. and Krylov, V. I., *Approximate methods of Higher Analysis*, translated by C. D. Benster, Groningen, P. Noordhoff Ltd., 1958.

27. Karp, S. N. and Karal, F. C., *Communs. pure appl. Math.* **15** (1962), 413.
28. Knopp, K., *Theory of Functions*, Parts 1 and 2, New York, Dover Publications, 1945.
29. Knowles, J. K., *Archs ration. Mech. Analysis*, **21** (1966), 1.
30. Kober, H., *Dictionary of Conformal Representations*, New York, Dover Publications, 1952.
31. Koiter, W. T., *Ing. Arch.* **28** (1959), 168.
32. Kolchanov, E. A. and Shvetsov, A. V., *Prik. Mekh.* **3** (1967), 46, to be translated in *Soviet Applied Mechanics* (Faraday Press, New York).
33. Kolosov, G. V., 'On the application of complex function theory to a plane problem of the mathematical theory of elasticity', Dorpat (Yuriev) University, 1909.
34. Kolosov, G. V. and Muskhelishvili, N. I., *Izv. Electrotechn. Inst. Petrograd*, **12** (1915), 39.
35. Lekhnitskii, S. G., *Theory of Elasticity of an Anisotropic Elastic Body*, translated by P. Fern, San Francisco, Holden Day, 1963.
36. Liebowitz, H. (editor), *Fracture*, New York, Academic Press, 1968.
37. Ling, C-B., *J. appl. Phys.* **19** (1948), 405.
38. Love, A. E. H., *Mathematical Theory of Elasticity*, 4th Edition, Cambridge, Cambridge University Press, 1927.
39. Mikhlin, S. G., *Integral Equations*, translated by A. H. Armstrong, London, Pergamon Press, 1957.
40. Milne-Thomson, L. M., *J. Lond. math. Soc.* **17** (1942), 115.
41. Milne-Thomson, L. M., *Plane Elastic Systems*, Berlin, Springer Verlag, 1960.
42. Mindlin, R. D., *Bull. Am. math. Soc.* **42** (1936), 373.
43. Muskhelishvili, N. I., *Some Basic Problems of the Mathematical Theory of Elasticity*, translated by J. R. M. Radok, Groningen, P .Noordhoff Ltd., 1963.
44. Muskhelishvili, N. I., *Singular Integral Equations*, translated by J. R. M. Radok, Groningen, P. Noordhoff, Ltd., 1953.
45. Nehari, Z., *Conformal Mapping*, New York, McGraw-Hill Book Co., 1952.
46. Neuber, H., *Theory of Notch Stresses*, translated by F. A. Raven, Ann Arbor (Michigan), Edwards Brothers, 1946. (or *Kerbspannungslehre*, 2nd edition, Berlin, Springer Verlag, 1958).
47. Nowacki, W., *Thermoelasticity*, Oxford, Pergamon Press, 1962.
48. Paris, P. C. and Sih, G. C. M., American Soc. for Testing and Materials, Publication 381 (1965), 30.
49. Rieder, G., *Abh. Braunschw. wiss. Ges.* **7** (1960), 4.
50. Savin, G. N., *Stress Concentrations around Holes* (translation edited by W. Johnson), London, Pergamon Press, 1961.
51. Schtaerman, I. Ya., *The Contact Problems of the Theory of Elasticity* (in Russian), Moscow-Leningrad, 1949.
52. Sherementev, M. P., *Prikl. Mat. Mekh.* **16** (1952), 437.
53. Sherman, D. I., *Trudӯ Seĭsm. Inst. A.H. SSSR*, **88** (1938).
54. Sneddon, I. N., *Fourier Transforms*, New York, McGraw-Hill Book Co., 1951.
55. Sneddon, I. N., *Mixed Boundary Value Problems in Potential Theory*, Amsterdam, North-Holland Publishing Co., 1966.

56. Sneddon, I. N., *Crack Problems in the Mathematical Theory of Elasticity*, Raleigh, North Carolina State College, 1961.
57. Sneddon, I. N. and Berry, D. S., 'The classical theory of elasticity', *Handbuch Der Physik*, Vol. VI, Berlin, Springer Verlag, 1958.
58. Sneddon, I. N. and Lowengrub, M., *Crack Problems in the Classical Theory of Elasticity*, S. I. A. M. monograph, New York, John Wiley & Sons, 1969.
59. Sokolnikoff, I. S., *Mathematical Theory of Elasticity*, New York, McGraw-Hill Book Co., 1956.
60. Sternberg, E., *Qt. appl. Math.* **11** (1954), 393.
61. Stevenson, A. C., *Proc. Roy. Soc. A*, **184** (1945), 129, 218.
62. Stippes, M., Wilson, H. B. and Krull, F. N., *Proc. 4th U.S. Natn. Congr. app. mech.*, A.S.M.E. **2** (1962), 799.
63. Teodorescu, P. P., 'One hundred years of investigations in the plane problem of the theory of elasticity', *App. Mech. Rev.* **17** (1964), 175.
64. Tiffen, R., *Q. Jl. Mech. app. Math.* **5** (1952), 237.
65. Tiffen, R. and Semple, H. M., *Mathematika* **12** (1965), 193.
66. Timoshenko, S. and Goodier, J. N., *Theory of Elasticity*, 2nd edition, New York, McGraw-Hill Book Co., 1951.
67. Titchmarsh, E. C., *Fourier Integrals*, London, Oxford University Press, 1948.
68. Toupin, R. A., *Archs ration. Mech. Analysis*, **18** (1965), 83
69. Truesdell, C. and Toupin, R. A., 'The classical field theories', *Handbuch Der Physik*, Vol 111/1, Berlin, Springer Verlag, 1960.
70. Truesdell, C. and Noll, W., 'The non-linear field theories of mechanics', *Handbuch Der Physik*, Vol 111/3, Berlin, Springer Verlag, 1965.
71. Volterra, V., *Ann. Éc. norm.* (*Sér* 3), **24** (1907), 401.
72. Westergaard, H. M., *J. appl. Mech.* **6** (1939), A49.
73. Williams, M. L., *J. appl. Mech.* **19** (1952), 526.
74. Williams, W. E., *Z.A.M.P.* **13** (1962), 133 and **14** (1963), 675.
75. Williams, W. E., *Proc. Glasg. math. Ass.* **6** (1964), 123.
76. Willmore, T. J., *Q. Jl. Mech. appl. Math.* **2** (1949), 53.
77. Wilson, H. B., *Developments in Theoretical and Applied Mechanics* **2** (1964), New York, Pergamon Press.

SUMMARY OF FORMULAE

The formulae below are normally referred to as the Kolosov-Muskhelishvili formulae.

The displacement and stress fields in Cartesian coordinates (equations (2.44))

$$\begin{aligned} 2\mu(u+\mathrm{i}v) = 2\mu D &= \kappa\Omega(z) - z\overline{\Omega'(z)} - \overline{\omega(z)}, \\ \tau_{xx}+\tau_{yy} &= 2\left\{\Omega'(z)+\overline{\Omega'(z)}\right\}, \\ \tau_{xx}-\tau_{yy}+2\mathrm{i}\tau_{xy} &= -2\left\{z\overline{\Omega''(z)}+\overline{\omega'(z)}\right\} \\ \tau_{yy}-\mathrm{i}\tau_{xy} &= \Omega'(z)+\overline{\Omega'(z)}+z\overline{\Omega''(z)}+\overline{\omega'(z)}. \end{aligned}$$

where $\kappa = (\lambda+3\mu)/(\lambda+\mu)$ for plane strain and $\kappa = (5\lambda+6\mu)/(3\lambda+2\mu)$ for generalised plane stress.

Resultant force $R(t)$ over an arc C acting on the material to the left as C is described (equation (2.76)).

$$\mathrm{i}R(t) = \left[\Omega(z) + z\overline{\Omega'(z)} + \overline{\omega(z)}\right]_{t_0}^{t} = \int_{t_0}^{t} (\tau_{\xi\xi}+\mathrm{i}\tau_{\xi\eta})\,dt.$$

Point force $X+\mathrm{i}Y$ and **moment** M acting at the point z_0 (equations (3.6)).

$$\begin{aligned} \Omega(z) &= -P\log(z-z_0) \\ \omega(z) &= \kappa\overline{P}\log(z-z_0) + \frac{P\overline{z}_0}{z-z_0} + \frac{\mathrm{i}M}{2\pi(z-z_0)} \end{aligned}$$

where $P = (X+\mathrm{i}Y)/[2\pi(1+\kappa)]$.

The displacement and stress fields in polar coordinates (equations (4.1))

$$\begin{aligned} 2\mu(u_r+\mathrm{i}u_\theta) &= 2\mu D\mathrm{e}^{-\mathrm{i}\theta} = e^{-\mathrm{i}\theta}\left\{\kappa\Omega(z) - z\overline{\Omega'(z)} - \overline{\omega(z)}\right\}, \\ \tau_{rr}+\tau_{\theta\theta} &= 2\left(\Omega'(z)+\overline{\Omega'(z)}\right) \\ \tau_{rr}-\tau_{\theta\theta}+2\mathrm{i}\tau_{r\theta} &= -2\left(z\overline{\Omega''(z)}+\overline{\omega'(z)}\right)\mathrm{e}^{-2\mathrm{i}\theta}, \\ \tau_{rr}+\mathrm{i}\tau_{r\theta} &= \Omega'(z)+\overline{\Omega'(z)}-\overline{z}\overline{\Omega''(z)}-(\overline{z}/z)\overline{\omega'(z)} \end{aligned}$$

ANSWERS TO THE EXAMPLES

Examples on Chapter 2 (p.46)

The first three examples give two further derivations of the general solution (2.44).

1. Use (2.31) and $2x_1 = z + \bar{z}$, $2ix_2 = z - \bar{z}$.
If $\theta_1 = \Theta(z) + \overline{\Theta(z)}$, $i\theta_2 = \psi(z) - \overline{\psi(z)}$, $i\phi_0 = \Phi(z) - \overline{\Phi(z)}$

then $\Omega(z) = \Theta(z) + \psi(z)$,

$$\omega(z) = z\left(\Theta'(z) - \psi'(z)\right) - \kappa\left(\Theta(z) - \psi(z)\right) - 2i\Phi(z).$$

2. Put
$$\begin{aligned}(\tau_{11}, \tau_{12}) &= (\phi_{,1} + \psi_{,2}\,, \quad \phi_{,2} - \psi_{,1})\\ (\tau_{21}, \tau_{22}) &= (\Phi_{,1} + \Psi_{,2}\,, \quad \Phi_{,2} - \Psi_{,1})\end{aligned}$$
Equilibrium implies $\nabla_1^2\phi = 0$, $\nabla_1^2\Phi = 0$.
Equality of τ_{12} terms implies $(\Phi + \psi)_{,1} + (-\phi + \Psi)_{,2} = 0$.
Put $\Phi + \psi = \alpha_{,1} + \beta_{,2}$, $-\phi + \Psi = \alpha_{,2} - \beta_{,1}$, then $\nabla_1^2\alpha = 0$.
Hence $\tau_{11}, \tau_{22}, \tau_{12}$ are given by (2.85) if we put $\beta = A$ and $\psi_{,12} = \phi_{,1} - \Phi_{,2} + \alpha_{,12}$.

Then $\psi_{,112} = \phi_{,11} - \Phi_{,12} + \alpha_{,112}$ so $\psi_{,11} = -\phi_{,2} - \Phi_{,1} + \alpha_{,12}$ using $\nabla_1^2\alpha = 0$.

3. $\tau_{11} + \tau_{22} = \nabla_1^2 A = \phi$, where $\nabla_1^2\phi = 0$. Put $\phi = 2\left\{\Omega'(z) + \overline{\Omega'(z)}\right\}$.

Then $\nabla_1^2 A = 4\dfrac{\partial^2 A}{\partial z \partial \bar{z}} = 2\left\{\Omega'(z) + \overline{\Omega'(z)}\right\}$
so $A = \frac{1}{2}\left(\bar{z}\Omega(z) + z\overline{\Omega(z)}\right) + \alpha(z) + \overline{\alpha(z)}$.
Similarly, since $\nabla_1^2\psi = 0$, $\psi = \beta(z) + \overline{\beta(z)}$.

$$\tau_{11} - \tau_{22} + 2i\tau_{12} = -\frac{4\partial^2}{\partial \overline{z}^2}(A + \mathrm{i}\psi) = -2\left(z\overline{\Omega''(z)} + \overline{\omega'(z)}\right)$$

where $\omega'(z) = 2\alpha''(z) + 2\mathrm{i}\beta''(z)$.

Integrate the second equation of (2.43) to give

$$2\mu D = -\left(z\overline{\Omega'(z)} + \overline{\omega(z)}\right) + \theta(z).$$

The first equation shows

$$\theta(z) = \kappa\Omega(z).$$

4. The stress boundary conditions reduce to (2.68) or (2.76). These equations determine $\Omega(z)$, $\omega(z)$ independently of the elastic constants for a simply-connected region. (For a multiply-connected region the requirement that the displacement field be single valued introduces elastic constants into the solution, see Example 9).

5. Rigid body displacement $\kappa\alpha - \overline{\beta}$. Rigid body rotation through a small angle $(\kappa + 1)A$.

6. $\mathrm{Re}(\alpha) = (X + Y)/4$, $\beta = (Y - X + 2\mathrm{i}S)/2$

On x = constant, $\tau_{\xi\xi} + \mathrm{i}\tau_{\xi\eta} = X + \mathrm{i}S$
On y = constant, $\tau_{\xi\xi} + \mathrm{i}\tau_{\xi\eta} = Y - \mathrm{i}S$
On $x + y$ = constant, $\tau_{\xi\xi} + \mathrm{i}\tau_{\xi\eta} = \frac{1}{2}(X+Y+2S) + \frac{\mathrm{i}}{2}(Y-X)$.

7. $\tau_{rr} + \mathrm{i}\tau_{r\theta} = n\left(\alpha z^{n-1} + \overline{\alpha}\,\overline{z}^{\,n-1}\right) - n(n-1)\overline{\alpha}\,\overline{z}^{\,n-1} - n\overline{\beta}\,\overline{z}^{\,n}/z$.

(a) $\tau_{rr} + \mathrm{i}\tau_{r\theta} = -p_2 + \mathrm{i}s_2$ on $|z| = b$.
No θ dependence so $n = 1$, $\beta = 0$ and $\alpha + \overline{\alpha} = -p_2 + \mathrm{i}s_2$.

Hence $s_2 = 0$, so there is no resultant moment on the disc.

[Solution $\Omega(z) = \alpha z$, $\omega(z) = 0$ where $\alpha + \overline{\alpha} = -p_2 + \mathrm{i}s_2$ and $s_2 = 0$ since the left-hand side is real.]

(b) Region $|z| > a$ with zero stress and rotation at infinity. Take $n = -1$

$$\tau_{rr} + \mathrm{i}\tau_{r\theta} = -\left(\frac{\alpha}{r^2}\mathrm{e}^{-2\mathrm{i}\theta} + \frac{\overline{\alpha}}{r^2}\mathrm{e}^{2\mathrm{i}\theta}\right) - \frac{2\overline{\alpha}}{r^2}\mathrm{e}^{2\mathrm{i}\theta} + \frac{\overline{\beta}}{r^2} = -p_1 + \mathrm{i}s_1$$

on $r = a$. No θ dependence hence $\alpha = 0$ and $\beta = -a^2(p_1 + \mathrm{i}s_1)$.
[Solution $\Omega(z) = 0$, $\omega(z) = \beta/z$ where $\beta = -a^2(p_1 + \mathrm{i}s_1)$]

(c) Solution $\Omega(z) = \alpha z$, $\omega(z) = \beta/z$ where

$$\begin{aligned}\beta\left(a^{-2} - b^{-2}\right) &= p_2 - p_1 + \mathrm{i}(s_2 - s_1)\\ (\alpha + \overline{\alpha})(a^2 - b^2) &= p_2 b^2 - p_1 a^2 + \mathrm{i}(s_1 a^2 - s_2 b^2)\end{aligned}$$

but the left-hand side is real, so $s_1 a^2 = s_2 b^2$, i.e. the resultant moment acting on the annulus must be zero.

8. $2\mu D = -\kappa P \log(|z|) + \overline{P}\mathrm{e}^{2\mathrm{i}\theta}$, where $P = \dfrac{X + \mathrm{i}Y}{2\pi(1+\kappa)}$.

Resultant force applied across an internal (clockwise) contour C surrounding the origin is

$$-\mathrm{i}\left[\Omega(z) + z\overline{\Omega'(z)} + \overline{\omega(z)}\right]_C = X + \mathrm{i}Y.$$

Using (2.74), the stress distribution on $r = a$ is

$$\tau_{rr} + \mathrm{i}\tau_{r\theta} = -\frac{X + \mathrm{i}Y}{2\pi a}\mathrm{e}^{-\mathrm{i}\theta} - \frac{(X - \mathrm{i}Y)}{\pi(1+\kappa)a}\mathrm{e}^{\mathrm{i}\theta}.$$

9. Use (2.74)

$$\tau_{rr} + \mathrm{i}\tau_{r\theta} = \Omega'(z) + \overline{\Omega'(z)} - \left(z\overline{\Omega''(z)} + \overline{\omega'(z)}\right)\overline{z}/z$$

to show

$$\tau_{rr} + \mathrm{i}\tau_{r\theta} = -(X + \mathrm{i}Y)\mathrm{e}^{-\mathrm{i}\theta}/(2\pi a).$$

10. Body-force potential due to a rotation with angular velocity ω about the origin is given by (2.46) to be $X(z, \overline{z}) = \frac{1}{2}\omega^2 \rho z^2 \overline{z}$.

The stress combinations Θ and Φ are given from (2.43) and (2.38) as

$$\begin{aligned}\Theta &= 2\left(\Omega'(z) + \overline{\Omega'(z)}\right) - \frac{\lambda+\mu}{\lambda+2\mu}\frac{\partial X}{\partial z}\\ \Phi &= -2\left(z\overline{\Omega''(z)} + \overline{\omega'(z)}\right) - \frac{\mu}{\lambda+2\mu}\frac{\partial X}{\partial \overline{z}}\end{aligned}$$

On the surface $|z| = a$ of the disc

$$\begin{aligned}\tau_{rr} + \mathrm{i}\tau_{r\theta} &= \tfrac{1}{2}\left[\Theta + \Phi \mathrm{e}^{-2\mathrm{i}\theta}\right]\\ &= \Omega'(z) + \overline{\Omega'(z)} - \left[\overline{z}\overline{\Omega''(z)} + \frac{\overline{z}}{z}\overline{\omega'(z)}\right] - \frac{\rho\omega^2 a^2(2\lambda + 3\mu)}{4(\lambda + 2\mu)}\end{aligned}$$

so the body force produces a uniform compression on the edge of the disc. This can be removed by adding the uniform tension field $\Omega(z) = \alpha z$, $\omega(z) = 0$ where $\alpha = \dfrac{2\lambda+3\mu}{8(\lambda+2\mu)}\rho\omega^2 a^2$. The combined radial stress becomes $\tau_{rr} = \frac{1}{4}\left(\dfrac{2\lambda+3\mu}{\lambda+2\mu}\right)\rho\omega^2(a^2-r^2)$ at radius r.

Note that for thin plates under generalised plane stress λ should be replaced by λ^*.

Examples on Chapter 3 (p.83)

1. Point forces: X at the origin, $-X$ at $x=a$, $y=0$.
Put $P = X/[2\pi(1+\kappa)]$
Potentials $\Omega(z) = -P(\log z - \log(z-a))$

$$\omega(z) = \kappa P(\log(z) - \log(z-a)) - P\frac{a}{z-a}.$$

Hence the displacement field is

$$2\mu D = -\frac{\kappa X}{\pi(1+\kappa)}(\log|z| - \log|z-a|) + \frac{X}{2\pi(1+\kappa)}\left(\frac{z}{\overline{z}} - \frac{z-a}{\overline{z}-a}\right)$$

and is bounded at infinity.

2. See §3.5. It is simplest to keep the manipulation in terms of $\Omega_0(z)$, $\omega_0(z)$ for as long as possible.
Displacement on $y=0$ is

$$\begin{aligned} 2\mu D &= (\kappa+1)\left[\Omega_0(x) - x\overline{\Omega_0'(x)} - \overline{\omega_0(x)}\right] \\ &= -\frac{Q}{2\pi}\log(x-z_0) - \frac{\kappa Q}{2\pi}\log(x-\overline{z}_0) + \frac{\overline{Q}}{2\pi}\frac{x-z_0}{x-\overline{z}_0}. \end{aligned}$$

where $Q = X + \mathrm{i}Y$.

3. Using $P = (X+\mathrm{i}Y)/[2\pi(1+\kappa)]$, τ_{yy} on $y=0$ has the value

$$2\mathrm{Re}\left\{P\left[\frac{\kappa}{x-\overline{z}_0} - \frac{1}{x-z_0} + \frac{z_0-\overline{z}_0}{(x-z_0)^2}\right]\right\}$$

4. Answer given.

5. The oscillatory regions are specified by the first zeros of $X'(z)$ as defined in (3.48) and (3.49).

6. The result follows immediately from (3.41) and (3.42).

7. Final result: $\Omega'(z) = -\frac{p_0}{2\pi \mathrm{i}} \log\left(\frac{z-a}{z+a}\right)$.

8. $\theta(z)$ satisfies, (see (3.58))

$$\begin{aligned} \theta'^{+}(x) + \theta'^{-}(x) &= \frac{4\mathrm{i}\mu f'(x)}{\kappa+1} \quad \text{on} \quad y=0,\ |x| \le a \\ \theta'^{+}(x) - \theta'^{-}(x) &= 0 \quad \text{on} \quad y=0,\ |x| \ge a. \end{aligned}$$

For a flat-ended punch with resultant force Y

$$\theta'(z) = -\frac{\mathrm{i}Y}{2\pi(z^2-a^2)^{\frac{1}{2}}}.$$

9. Suppose the contact region is of length $2a$ and put $x = \xi + a$ so that the contact region is $y=0,\ |\xi| \le a$.

The corresponding complex potential is

$$\Omega'(z) = -\frac{\mathrm{i}}{2\pi}\left(\frac{Y}{(z^2-a^2)^{\frac{1}{2}}} + \frac{4\pi\mu\varepsilon}{1+\kappa}\left[1 - \frac{z}{(z^2-a^2)^{\frac{1}{2}}}\right]\right)$$

where $z = \xi + \mathrm{i}y$.

The contact stress τ_{yy} is

$$\begin{aligned} \tau_{yy} &= -\frac{1}{\pi\sqrt{a^2-\xi^2}}\left(Y - \frac{4\pi\mu\varepsilon\xi}{1+\kappa}\right) \\ &= -\frac{1}{\pi\sqrt{x(2a-x)}}\left(Y + \frac{4\pi\mu\varepsilon}{1+\kappa}(a-x)\right), \quad 0 \le x \le 2a \end{aligned}$$

length of the contact region $0 \le x \le 2a$ is given by

$$a = (1+\kappa)Y/4\pi\mu\varepsilon.$$

The solution holds until $2a = l$, the width of the punch or, equivalently, for $0 \le Y \le 2\pi\mu\varepsilon l/(1+\kappa)$.

10. Answer is given.

11. The representation is found by using the displacement continuation to define $\phi(z)$ and the stress continuation to define $\theta(z)$.

12. The functions $\theta(z)$ and $\phi(z)$ are holomorphic in the whole plane so that $\theta(z) = \alpha z$, $\phi(z) = \beta z$. A solution exists only if

$$\begin{aligned} Y_1 &= Y_2, \\ X_2\mu_1(1+\kappa_2) - X_1\mu_2(1+\kappa_1) &= Y_1\left(3(\mu_1-\mu_2) + \mu_2\kappa_1 - \mu_1\kappa_2\right). \end{aligned}$$

These relations ensure that the displacements on $y = 0\pm$ namely

$$D_1 = [(\kappa_1 + 1)X_1 + (\kappa_1 - 3)Y_1]\, x/8\mu_1,$$
$$D_2 = [(\kappa_2 + 1)X_2 + (\kappa_2 - 3)Y_2]\, x/8\mu_2,$$

are compatible.

13. Using the representation in example 11, show that on the upper and lower faces of the crack

$$\text{as } y \to 0^+, \quad \tau_{yy} - \mathrm{i}\tau_{xy} = \frac{\mu_1\theta'^+ + \phi'^+}{\mu_1 + \mu_2\kappa_1} + \frac{\mu_2\theta'^- + \phi'^-}{\mu_2 + \mu_1\kappa_2} - \theta'^- = -p(x)$$

$$\text{as } y \to 0^-, \quad \tau_{yy} - \mathrm{i}\tau_{xy} = \frac{\mu_1\theta'^+ + \phi'^+}{\mu_1 + \mu_2\kappa_1} + \frac{\mu_2\theta'^- + \phi'^-}{\mu_2 + \mu_1\kappa_2} - \theta'^+ = -p(x).$$

Since the resultant force applied to the crack is zero, $\theta'(z)$ and $\phi'(z)$ have $O(1/z^2)$ as $|z| \to \infty$. Hence $\theta(z) \equiv 0$, and from (1.33) and (1.34) (since $\kappa = -\alpha$ and $\gamma = \frac{1}{2} - \mathrm{i}\beta$)

$$\phi'(z) = \frac{X(z)}{2\pi\mathrm{i}} \int_L \frac{-(\mu_1 + \mu_2\kappa_1)p(t)\mathrm{d}t}{X^+(t)(t - z)} + X(z)P(z)$$

where $X(z) = \dfrac{1}{(z^2 - a^2)^{\frac{1}{2}}} \left(\dfrac{z + a}{z - a}\right)^{\mathrm{i}\beta}$ and $2\pi\beta = \log\alpha$.

When $p(t)$ is the constant p_0, the integral may be evaluated using §1.8 and the result (1.43), to give

$$\phi'(z) = K\left[1 - (z - 2\mathrm{i}a\beta)\left(\frac{z + a}{z - a}\right)^{\mathrm{i}\beta} \frac{1}{(z^2 - a^2)^{\frac{1}{2}}}\right]$$

where $K = -\dfrac{(\mu_1 + \mu_2\kappa_1)}{1 + \alpha} p_0$.

Hence $\phi(z)$ may be found on integration.

The normal distance apart of the crack surface is proportional to $\mathrm{Im}(\phi^+ - \phi^-)$ on $y = 0$, $|x| \le a$, yielding

$$\frac{p_0}{2\mu_1\mu_2}\sqrt{(\mu_1 + \mu_2\kappa_1)(\mu_2 + \mu_1\kappa_2)}(a^2 - x^2)^{\frac{1}{2}} \cos\left(\beta \log\left|\frac{x + a}{x - a}\right|\right),$$

on noting $\arg\left(\dfrac{z + a}{z - a}\right) = \mp\pi$ on $y = 0\pm$, $|x| \le a$.

The physical implications of this 'crossover' phenomenon in linear elasticity have been investigated by several authors since it was first discovered in the 1960's, see reference 10.

Examples on Chapter 4 (p.125)

1. and **2.** Answers given.

3. New radius of disc $= R + \delta_1$, new radius of hole $= a + \delta_2$. Then $R + \delta_1 = a + \delta_2$. Stress on the interface must be continuous so that $4\mu_1\delta_1/\{(\kappa_1 - 1)R\} = -2\mu_2\delta_2/a$. Hence the pressure on the interface is

$$-\tau_{rr} = \frac{4\mu_1\mu_2}{\mu_2(\kappa_1 - 1)(R/a) + 2\mu_1}\left(\frac{R}{a} - 1\right)$$

and its radius is

$$R - \frac{(R-a)\mu_2(\kappa_1 - 1)}{\mu_2(\kappa_1 - 1) + 2\mu_1(a/R)}.$$

To first-order terms in $(R/a - 1)$ these denominators simplify to $\mu_2(\kappa_1 - 1) + 2\mu_1$.

4. There must be a resultant force $X + \mathrm{i}Y$ over the 'hole' $|z| = a$, so $\Omega(z)$ and $\omega(z)$ are given by (4.3). To satisfy (4.4) it is sufficient to put $\Omega_0(z) = \alpha z^2 + \beta$ and $\omega_0(z) = \gamma/z^2$. The boundary conditions determine α, β, γ and show the resultant force has the real value given in the question.

5.
$$\begin{aligned} F(t) &= \int_{t_0}^{t} (\tau_{rr} + \mathrm{i}\tau_{r\theta})dt \\ &= -\frac{p}{2a\alpha}\begin{cases} t - a\mathrm{e}^{-\mathrm{i}\alpha}, & -\alpha \le \arg t \le \alpha \\ a(\mathrm{e}^{\mathrm{i}\alpha} - \mathrm{e}^{-\mathrm{i}\alpha}), & \alpha \le \arg t \le \pi - \alpha \\ a\mathrm{e}^{\mathrm{i}\alpha} + t, & \pi - \alpha \le \arg t \le \pi + \alpha \\ 0, & \pi + \alpha \le \arg t \le 2\pi - \alpha. \end{cases} \end{aligned}$$

Hence from (4.37), $\alpha_1 + \overline{\alpha}_1 = \dfrac{1}{2\pi\mathrm{i}}\displaystyle\int_\Gamma \frac{F(t)dt}{t^2} = -\frac{p}{\pi a}$, $\alpha_2 = 0$.

From (4.36), $\Omega'(z) + \alpha_1 = \dfrac{1}{2\pi\mathrm{i}}\displaystyle\int_\Gamma \frac{F(t)dt}{(t-z)^2}$.

The integral needs to be performed in sections as the circle $|t| = a$ is described in an anti-clockwise direction.

6. Exterior region $|z| \geq a$. The contour Γ is the circle $|z| = a$ described in a clockwise direction. The resultant force $X + \mathrm{i}Y$ acts over the hole so $\Omega(z) = -P\log(z) + \Omega_0(z)$, $\omega(z) = \kappa\overline{P}\log(z) + \omega_0(z)$ where $P = (X+\mathrm{i}Y)/[2\pi(1+\kappa)]$. Since $\Omega(z)$ and $\omega(z)$ are undefined to within a constant, we can assume $\Omega_0(z)$, $\omega_0(z)$ tend to zero as $|z| \to \infty$. On $t = a\mathrm{e}^{\mathrm{i}\theta}$,

$$\begin{aligned}\Omega_0(t) + t\overline{\Omega_0'(t)} + \overline{\omega_0(t)} &= F(t)\\ &= \mathrm{i}R(t) + P(\log t - \kappa \log \bar{t}) + \overline{P}t/\bar{t}\\ &= \mathrm{i}R(t) + (X+\mathrm{i}Y)\mathrm{i}\theta/2\pi + \overline{P}\mathrm{e}^{2\mathrm{i}\theta} + \text{ constant}\end{aligned}$$

where $\mathrm{i}R(t) = \displaystyle\int_0^t (\tau_{rr} + \mathrm{i}\tau_{r\theta})dt = -(X+\mathrm{i}Y)\mathrm{i}\theta/2\pi$.

Hence $F(t) = \overline{P}\mathrm{e}^{2\mathrm{i}\theta} +$ constant.

$$\Omega_0(z) - \Omega_0(\infty) = \frac{1}{2\pi\mathrm{i}}\int_\Gamma \frac{\Omega_0(t)}{t-z}dt = \frac{1}{2\pi\mathrm{i}}\int_\Gamma \frac{F(t) - t\,\overline{\Omega_0'(t)} - \overline{\omega_0(t)}dt}{t-z}.$$

Hence $\Omega_0(z) = 0$ since $\Omega_0(\infty) = 0$, $\omega_0(\infty) = 0$, $\displaystyle\int_\Gamma \frac{t^2}{t-z}dt = 0$, $z \in S^+$.

$$\begin{aligned}\text{Similarly } \omega_0(z) - \omega_0(\infty) &= \frac{1}{2\pi\mathrm{i}}\int_\Gamma \frac{\overline{F(t)} - \overline{\Omega_0(t)} - \bar{t}\Omega_0'(t)dt}{t-z}\\ \omega_0(z) &= \frac{P}{2\pi\mathrm{i}}\int_\Gamma \frac{a^2}{t^2}\frac{1}{t-z}dt = \frac{X+\mathrm{i}Y}{2\pi(1+\kappa)}\frac{a^2}{z^2}.\end{aligned}$$

7. Behaviour at infinity $\Omega(z) = \frac{T}{4}z + \Omega_0(z)$, $\omega(z) = -\frac{T}{2}z + \omega_0(z)$. No resultant force applied over the boundary, so $\Omega_0(z) = \alpha_0 + O(\frac{1}{z})$, $\omega_0(z) = \beta_0 + O(\frac{1}{z})$ as $|z| \to \infty$. On $|z| = a$

$$\kappa\Omega_0(z) - z\overline{\Omega_0'(z)} - \overline{\omega_0(z)} = -\tfrac{1}{4}(\kappa - 1)Tz - \tfrac{1}{2}T\bar{z}$$

$$\begin{aligned}\Omega_0(z) - \alpha_0 &= \frac{1}{2\pi\mathrm{i}\kappa}\int_\Gamma \frac{t\overline{\Omega_0'(t)} + \overline{\omega_0(t)} - \frac{1}{4}(\kappa-1)Tt - \frac{1}{2}Ta^2/t}{t-z}dt\\ &= -\frac{Ta^2}{2\kappa z},\end{aligned}$$

where the contour Γ is $t\bar{t} = a^2$ described in a clockwise direction.

$$\begin{aligned}\omega_0(z) - \beta_0 &= \frac{1}{2\pi \mathrm{i}} \int_\Gamma \frac{\kappa\overline{\Omega_0(t)} - \bar{t}\Omega_0'(t) + \frac{1}{4}(\kappa-1)Ta^2/t + \frac{1}{2}Tt}{t-z} dt \\ &= -\frac{1}{2\pi \mathrm{i}} \int_\Gamma \frac{Ta^4}{2\kappa} \frac{dt}{t^3(t-z)} + \frac{1}{4}(\kappa-1)\frac{Ta^2}{z} \\ &= \frac{1}{4}(\kappa-1)\frac{Ta^2}{z} - \frac{Ta^4}{2\kappa z^3}.\end{aligned}$$

Hence $\Omega'(z)$ and $\omega'(z)$.

8. Stress continuation implies $\Omega'^+(t) - \Omega'^-(t) = \tau_{rr} + \mathrm{i}\tau_{r\theta}$ on $|t| = a$. Hence using the Cauchy integral (4.54),

$$\Omega'(z) = \frac{1}{2\pi \mathrm{i}} \int_L \left(-\frac{p}{2a\alpha}\right) \frac{1}{t-z} dt + \text{constant}$$

where the line integral is taken over the two arcs on which the pressure is applied.

9. Displacement continuation using $\omega(z) = \kappa\overline{\Omega(a^2/\bar{z})} - (a^2/z)\Omega'(z)$ shows that $\Omega^+(z) - \Omega^-(z) = 0$ on $|z| = a$. Hence $\Omega(z)$ is analytic in the whole plane except at $z = 0$ and infinity.

Hence $\Omega(z) = \dfrac{T}{4}\left(z - \dfrac{2a^2}{\kappa z}\right)$.

10. The solution follows from (4.64).

$$\Omega_0(z) = P\log\left(\frac{z-b}{z}\right), \quad \omega_0(z) = -\kappa P \log\left(\frac{z-b}{z}\right) - \frac{Pb}{z-b},$$

$$P = \frac{X}{2\pi(1+\kappa)}.$$

On $|z| = a$, with $b = \frac{1}{2}a$.

$$\tau_{\theta\theta} = 4P\mathrm{Re}\left[-\frac{1}{z} + \frac{1}{z-b} + \frac{2z}{a^2} + \frac{1}{b} + (\kappa-1)\frac{b}{2a^2} - \frac{\kappa b}{a^2 - bz} + \frac{a^2(b^2-a^2)}{b(a^2-bz)^2}\right].$$

11. $$\Omega'(z) = -\frac{P}{z} + \frac{2Pz}{a^2} + (1+\kappa)P\left[\frac{1}{2a} + \frac{1}{z-a}\right]$$

$$\omega'(z) = \frac{\kappa P}{z} + (1+\kappa)P\left[\frac{a}{(z-a)^2} - \frac{1}{z-a}\right]$$

where $P = X/[2\pi(1+\kappa)]$.

12. $\Omega(z)$ and $\omega(z)$ are defined for $|z| \geq a$.

$$\Omega(z) \to -P\log z,\ \omega(z) \to \kappa\overline{P}\log z + \mathrm{i}M/(2\pi z) \quad \text{as } |z| \to \infty$$

where $P = \dfrac{X+\mathrm{i}Y}{2\pi(1+\kappa)}$.

Displacement continuation $\Omega(z) = \frac{1}{\kappa}\left(z\overline{\Omega'(a^2/\overline{z})} + \overline{\omega(a^2/\overline{z})}\right)$ when $|z| < a$.

$$\Omega'(z) \to -\frac{P}{z} - \frac{\mathrm{i}M}{2\pi\kappa a^2} \text{ for small } |z|.$$

Using the displacement continuation

$$\Omega^+(t) - \Omega^-(t) = \frac{2\mu}{\kappa}(u_0 + \mathrm{i}v_0 + \mathrm{i}\varepsilon t) \text{ on } |t| = a.$$

Hence

$$\Omega(z) = \begin{cases} \phi(z), & |z| > a \\ -\dfrac{2\mu}{\kappa}(u_0 + \mathrm{i}v_o + \mathrm{i}\varepsilon z) + \phi(z), & |z| < a. \end{cases}$$

where $\phi(z)$ is analytic except at the origin and infinity. Behaviour of $\Omega(z)$ at infinity implies $\phi(z) = -P\log z$, behaviour at the origin shows $M = 4\pi a^2\mu\varepsilon$. Hence

$$\begin{aligned} \Omega'(z) = -P/z,\ \omega'(z) &= \frac{d}{dz}\left[\kappa\overline{\Omega(a^2/\overline{z})} - (a^2/z)\,\Omega'(z)\right] && \text{for } |z| \geq a \\ \omega'(z) &= \kappa\overline{P}/z - \mathrm{i}M/(2\pi z^2) - 2a^2P/z^3 && \text{for } |z| \geq a. \end{aligned}$$

13. Denote the resultant force applied over $|z| = a$ by $X + \mathrm{i}Y$. Then $\Omega(z) = -P\log z + \Omega_0(z)$, $\omega(z) = \kappa\overline{P}\log z + \omega_0(z)$ where $P = (X+\mathrm{i}Y)/[2\pi(1+\kappa)]$ and $\Omega_0(z)$, $\omega_0(z)$ are single-valued in the annulus. Then

$$\kappa\Omega_0(z) - z\overline{\Omega_0'(z)} - \overline{\omega_0(z)} = 2\kappa P\log(|z|) - \overline{P}z/\overline{z} + 2\mu D(t)$$

on $|z| = a$ and $|z| = b$.

Use stress continuation to show

$$\Omega_0(z) = 2P\log b - \frac{2Pb^2}{\kappa^2(a^2+b^2)} - \frac{Pz^2}{\kappa(a^2+b^2)},\quad \omega_0(z) = \frac{Pa^2b^2}{(a^2+b^2)}\frac{1}{z^2}$$

in $a \le |z| \le b$ where $X + \mathrm{i}Y$ and hence P are defined in Q4 and the constant in $\omega_0(z)$ has been set to zero. Stress on $|z| = b$ is

$$\begin{aligned}\tau_{rr} + \mathrm{i}\tau_{r\theta} &= -\frac{2P}{b}\left(1+\frac{2b^2}{\kappa(a^2+b^2)}\right)\cos\theta + \frac{P}{b}\left(\frac{a^2-b^2}{a^2+b^2}\right)\mathrm{e}^{\mathrm{i}\theta} \\ &\quad + \frac{P}{b}\left(\frac{2b^2}{\kappa(a^2+b^2)} - \kappa\right)\mathrm{e}^{-\mathrm{i}\theta}.\end{aligned}$$

14. From (4.81), $\Omega'(z) = \frac{T}{4} f(z)X(z)$ where $f(z)$ is defined by the term in square brackets and $X(z)$ is the Plemelj function $(z - a\mathrm{e}^{\mathrm{i}\phi})^{-\gamma}(z - a\mathrm{e}^{-\mathrm{i}\phi})^{\gamma-1}$ where $\gamma = \frac{1}{2} + \mathrm{i}\beta$.
Since $\kappa X^+(z) + X^-(z) = 0$ on the arc L, the stress field

$$\tau_{rr} + \mathrm{i}\tau_{r\theta} = \Omega'^+(z) - \Omega'^-(z) = \frac{T}{4}(1+\kappa)f(z)X^+(z) \text{ on } L.$$

Near the end $a\mathrm{e}^{\mathrm{i}\phi}$, $z = a\mathrm{e}^{\mathrm{i}\phi} + \zeta$ where $\zeta = s\mathrm{e}^{\mathrm{i}(\phi-\pi/2)}$ and s is a small arclength. Then

$$\begin{aligned}X^+(z) &= \zeta^{-\frac{1}{2}-\mathrm{i}\beta}(2\mathrm{i}a\sin\phi + \zeta)^{-\frac{1}{2}+\mathrm{i}\beta} \\ &= s^{-\frac{1}{2}}\mathrm{e}^{-\mathrm{i}\beta\log(s)}\mathrm{e}^{(\beta-\mathrm{i}/2)(\phi-\pi/2)}(2\mathrm{i}a\sin\phi)^{-\frac{1}{2}-\mathrm{i}\beta}.\end{aligned}$$

to leading order in s on the arc L. Hence there is an oscillating stress singularity at the end $a\mathrm{e}^{\mathrm{i}\phi}$. Similarly there is an oscillating singularity in $\frac{\partial D}{\partial\theta}$ on $|z| = a$ as $a\mathrm{e}^{\mathrm{i}\phi}$ is approached along the unreinforced surface.

15. The result follows directly from §4.6.

16. From (4.90), $\theta'(z) = -\overline{\Omega'(0)}$.
From (4.91) and (1.43)

$$2\Omega'(z) - \theta'(z) = -p\left(1 - \frac{z}{\sqrt{z^2+a^2}}\right) + \frac{(D_0 + D_1 z)}{\sqrt{z^2+a^2}}$$

since $\beta_2 = 0$, $\beta_1 = 0$ from the behaviour of $\Omega'(z)$ at $z = 0$.
Behaviour at infinity shows $D_1 = \overline{\Omega'(0)}$ and $D_0 = 0$.
At $z = 0$, $\quad 2\Omega'(0) + \overline{\Omega'(0)} = -p$.
Hence $\Omega'(z) = -\dfrac{p}{3}\left(1 - \dfrac{z}{\sqrt{z^2+a^2}}\right)$.
An infinite plane under an all-round tension T generates a uniform tension T over the line of the crack. This tension may be removed by applying the pressure T to the crack, hence

$$\Omega'(z) = \frac{T}{3}\frac{z}{\sqrt{z^2+a^2}} + \frac{T}{6}.$$

17. Solution $\theta(z) = Az$, $\phi(z) = Bz + CX(z)$ where

$$X(z) = (z - ae^{i\phi})^{\frac{1}{2}+i\beta}(z - ae^{-i\phi})^{\frac{1}{2}-i\beta}.$$

Boundary conditions on crack

$$\mu_1(1 - \alpha\kappa_2)A + (1 + \alpha)B = -\tfrac{1}{2}(\mu_1 + \mu_2\kappa_1)p.$$

Condition at infinity: $\mu_1 A + B + C = 0$.
Condition at origin: $A(2\mu_2 + \mu_1\kappa_2) + B + CX'(0) = 0$.
Hence

$$A\,[2\mu_2(1+\kappa_2)+\mu_1(1+\kappa_2)-\alpha\mu_1(1+\kappa_2)X'(0)]=\tfrac{1}{2}(\mu_1+\mu_2\kappa_1)p(1-X'(0))$$

where $X'(0) = e^{-2\beta\phi}(\cos\phi + 2\beta\sin\phi)$.
The displacement $D_1 = u_{1r} + iu_{1\theta}$ on the outer crack surface is given by

$$2\mu_1 e^{i\theta} D_1 = \frac{1}{\mu_1 + \mu_2\kappa_1}\left[\mu_1(\kappa_1 + \alpha\kappa_2)Az + \kappa_1\phi^+ - \alpha\phi^-\right]$$

hence $D_1 - D_2$ may be found. Near the end of the crack the distance apart of its surfaces depends on $s^{\frac{1}{2}}e^{i\beta\log(s)}$ as $s \to 0$, where $s = |e^{i\theta} - e^{\pm i\phi}|$.

18. $\kappa\Omega(z) - z\overline{\Omega'(z)} - \overline{\omega(z)} = 2\mu\varepsilon\sin(n\theta)e^{i\theta}$ on $|z| = a$.

$$\Omega(z) = i\frac{\mu\varepsilon}{\kappa}\left(\frac{a}{z}\right)^{n-1}, \quad \omega(z) = -i\mu\varepsilon\left(1 + \frac{1-n}{\kappa}\right)\left(\frac{a}{z}\right)^{n+1}.$$

19. $\Omega(z) = \frac{1}{4}(N_1+N_2)z+\frac{1}{2}(N_1-N_2)e^{2i\alpha}a^2/z$
$\omega(z) = -\frac{1}{2}(N_1-N_2)e^{-2i\alpha}z-\frac{1}{2}(N_1+N_2)a^2/z+\frac{1}{2}(N_1-N_2)e^{2i\alpha}a^4/z^3$.
On $z = ae^{i\theta}$, $\tau_{\theta\theta} = N_1 + N_2 - 2(N_1 - N_2)\cos 2(\alpha - \theta)$.
Max/min values at $\theta = \alpha$, $\alpha + \pi$, $\alpha \pm \pi/2$.

20. $\Omega_0(z) = -P\log z + O(z^{-1})$, $\omega_0(z) = \kappa\overline{P}\log z + \beta_1/z + O(z^{-2})$ as $|z| \to \infty$ where $P = \sum(X_k + iY_k)/\left[2\pi(1+\kappa)\right]$
$\Omega(z) = \Omega_0(z) + \Omega_1(z)$, $\omega(z) = \omega_0(z) + \omega_1(z)$

$\kappa\Omega_1(z) = z\overline{\Omega_0'(a^2/\overline{z})} + \overline{\omega_0(a^2/\overline{z})} + \kappa P\log z + Az$

$\omega_1(z) = \kappa\overline{\Omega_0(a^2/\overline{z})} - (a^2/z)\Omega_1'(z) + \kappa\overline{P}\log(a^2/z) + \overline{A}a^2/z$

where $\kappa\overline{A} - A = \overline{\beta}_1/a^2$.

With point force X at the origin $P = X/[2\pi(1+\kappa)]$

$$\Omega(z) = -P\log(z/a^2) - Pz^2/(\kappa a^2),\ \omega(z) = \kappa\overline{P}\log z + 2\overline{P}/\kappa.$$

On the boundary $|z| = a$.

$$\tau_{rr} + \mathrm{i}\tau_{r\theta} = -\frac{P}{a}\left[\left(3 + \frac{2}{\kappa} + \kappa\right)\cos\theta + \mathrm{i}\left(1 + \frac{2}{\kappa} - \kappa\right)\sin\theta\right].$$

21. Suppose the insert is displaced a distance d under the force X. Boundary conditions are $D = 0$ on $|z| = b$, $u_r = \varepsilon + d\cos\theta$ and $\tau_{r\theta} = 0$ on $|z| = a$.
Complex potentials

$$\begin{aligned}\Omega(z) &= -P\log z + \alpha_0 + \alpha_1 z + \alpha_2 z^2\\ \omega(z) &= \kappa\overline{P}\log z + \beta_1/z + \beta_2/z^2\end{aligned}$$

where $P = X/\left[2\pi(1+\kappa)\right]$.

$$\begin{aligned}\text{Hence } \alpha_1 &= \frac{-2\mu a\varepsilon}{(\kappa-1)(b^2-a^2)},\ \beta_1 = (\kappa-1)b^2\alpha_1\\ \alpha_2 &= -P(b^2 + \tfrac{1}{2}(\kappa-1)a^2)/(a^4+\kappa b^4),\ \beta_2 = Pb^2 + \kappa b^4\alpha_2\\ \kappa\alpha_0 &= 2b^2\alpha_2 + \kappa P\log(b^2)\\ d &= \tfrac{1}{2}(\kappa+1)P + \kappa P\log\left(b^2/a^2\right) - \tfrac{1}{2}P\frac{\left((\kappa-1)a^2 + 2b^2\right)^2}{a^4+\kappa b^4}\end{aligned}$$

On the circle $|z| = a$ separation occurs at $\theta = \pi$ when

$$X = 2\pi\mu\varepsilon\left[\frac{2a^2 + (\kappa-1)b^2}{(\kappa-1)(b^2-a^2)}\right].$$

Examples on Chapter 5 (p.170)

The questions **1**, **3**, **4**, **5** and **11** contain the answer as part of the question.

2. With a uniform tension T, the complete potentials are

$$\begin{aligned}\Omega(\zeta) &= \frac{RT}{4}\zeta + \left[8\mu\mathrm{i}\varepsilon Rm + RTm - 2RT\mathrm{e}^{2\mathrm{i}\alpha}\right]\frac{1}{4\kappa\zeta}\\ \omega(\zeta) &= -\frac{RT}{2}\mathrm{e}^{-2\mathrm{i}\alpha}\zeta + \left[8\mu\mathrm{i}\varepsilon R + RT(\kappa-1)\right]\frac{1}{4\zeta}\\ &\quad + \frac{1+m\zeta^2}{4\kappa\zeta\left(\zeta^2-m\right)}\left[8\mu\mathrm{i}\varepsilon Rm - RT(\kappa-1)m - 2RT\mathrm{e}^{2\mathrm{i}\alpha}\right]\end{aligned}$$

The zero moment condition is found from using (2.48) and calculating the term in $1/\zeta$ in $\omega(\zeta)(1-m/\zeta^2)$.

6. The effect of the point force X is found by taking the limit of $\Omega(\zeta)$ as $\theta_0 \to 0$ where $p = X/\left[2R(1-m)\sin\theta_0\right]$.

7. Point forces and moments may be represented by the leading-order terms at the singularities, so that $\log(z) \to \log(m(\zeta)) \to \log(R\zeta)$, $1/z \to 1/m(\zeta) \to 1/R\zeta$, with the additional terms being absorbed into $\Omega_1(\zeta)$ and $\omega_1(\zeta)$.

$$\Omega(\zeta) = -\frac{iM}{2\pi R\kappa}\zeta - \frac{1}{\kappa}\left(\beta_1\zeta + \beta_2\zeta^2\right)$$

$$\omega(\zeta) = \frac{iM}{2\pi R\zeta} + (2m^2-\kappa)\frac{2\beta_2}{\kappa} + \frac{2m(2m^2-1)}{(1+2m\zeta)}\frac{iM}{2\pi R\kappa}\left(1+\frac{1}{\kappa}-\frac{1}{\kappa-2m^2}\right)$$

where $(1+\kappa)(\kappa-2m^2)\beta_1 = \dfrac{iM}{2\pi R}\left(-\kappa+2m^2(1+\kappa)\right)$

$$(1+\kappa)(\kappa-2m^2)\beta_2 = \frac{iMm}{2\pi R}\left\{1-(1+\kappa)\left(1-2m^2/\kappa\right)\right\} + \frac{iMm^3}{\pi R}(1+\kappa).$$

8. $\Omega(\zeta) = \dfrac{RT}{2}\left(\zeta - \dfrac{m}{\gamma\zeta}\right) +$ constant where $\gamma = 1$ when the hole is unstressed, $\gamma = -\kappa$ when $D = 0$.
When $D = 0$, the stress concentration factors are
$\dfrac{T}{b\kappa}\left[a(\kappa-1)+b(\kappa+1)\right]$, $\dfrac{T}{a\kappa}\left[a(\kappa+1)+b(\kappa-1)\right]$.

9.
$$\Omega(\zeta) = \frac{RT}{4}\zeta + \tfrac{1}{2}RTe^{2i\alpha}\frac{1}{(1-3a_2^2)}\frac{1}{\zeta} - \frac{RT}{4}\left(\frac{1+2a_2}{1-2a_2}\right)\frac{a_1}{\zeta^2}$$
$$+\tfrac{1}{2}RTe^{-2i\alpha}\frac{a_2}{(1-3a_2^2)}\frac{1}{\zeta^3} - \frac{RT}{4}\frac{a_2}{\zeta^5} + \text{constant}.$$

Hoop stress $\tau_{\eta\eta} = 4\mathrm{Re}\left(\dfrac{\Omega'(\zeta)}{m'(\zeta)}\right)$ to be evaluated at $\zeta = 1$.

10. See answers to 2,4,8 and 9.

12. Put $e^{i\eta} = e^{i\phi} + i\varepsilon$, then $|R_1| = |\varepsilon|$, $|R_2| = |2\sin\phi + \varepsilon|$.
Oscillation near $e^{i\phi}$ commences when
$\beta\log|2\sin\phi/\varepsilon| = \pi/2$, so $|\varepsilon| = |2\sin\phi|e^{-\pi/(2\beta)}$.

AUTHOR INDEX

SUBJECT INDEX

A CATALOG OF SELECTED

DOVER BOOKS

IN SCIENCE AND MATHEMATICS

A CATALOG OF SELECTED

DOVER BOOKS

IN SCIENCE AND MATHEMATICS

Astronomy

BURNHAM'S CELESTIAL HANDBOOK, Robert Burnham, Jr. Thorough guide to the stars beyond our solar system. Exhaustive treatment. Alphabetical by constellation: Andromeda to Cetus in Vol. 1; Chamaeleon to Orion in Vol. 2; and Pavo to Vulpecula in Vol. 3. Hundreds of illustrations. Index in Vol. 3. 2,000pp. 6⅛ x 9¼.
23567-X, 23568-8, 23673-0 Three-vol. set

THE EXTRATERRESTRIAL LIFE DEBATE, 1750–1900, Michael J. Crowe. First detailed, scholarly study in English of the many ideas that developed from 1750 to 1900 regarding the existence of intelligent extraterrestrial life. Examines ideas of Kant, Herschel, Voltaire, Percival Lowell, many other scientists and thinkers. 16 illustrations. 704pp. 5⅜ x 8½. 40675-X

A HISTORY OF ASTRONOMY, A. Pannekoek. Well-balanced, carefully reasoned study covers such topics as Ptolemaic theory, work of Copernicus, Kepler, Newton, Eddington's work on stars, much more. Illustrated. References. 521pp. 5⅜ x 8½.
65994-1

AMATEUR ASTRONOMER'S HANDBOOK, J. B. Sidgwick. Timeless, comprehensive coverage of telescopes, mirrors, lenses, mountings, telescope drives, micrometers, spectroscopes, more. 189 illustrations. 576pp. 5⅝ x 8¼. (Available in U.S. only.)
24034-7

STARS AND RELATIVITY, Ya. B. Zel'dovich and I. D. Novikov. Vol. 1 of *Relativistic Astrophysics* by famed Russian scientists. General relativity, properties of matter under astrophysical conditions, stars, and stellar systems. Deep physical insights, clear presentation. 1971 edition. References. 544pp. 5⅝ x 8¼. 69424-0

Chemistry

CHEMICAL MAGIC, Leonard A. Ford. Second Edition, Revised by E. Winston Grundmeier. Over 100 unusual stunts demonstrating cold fire, dust explosions, much more. Text explains scientific principles and stresses safety precautions. 128pp. 5⅜ x 8½. 67628-5

THE DEVELOPMENT OF MODERN CHEMISTRY, Aaron J. Ihde. Authoritative history of chemistry from ancient Greek theory to 20th-century innovation. Covers major chemists and their discoveries. 209 illustrations. 14 tables. Bibliographies. Indices. Appendices. 851pp. 5⅜ x 8½. 64235-6

CATALYSIS IN CHEMISTRY AND ENZYMOLOGY, William P. Jencks. Exceptionally clear coverage of mechanisms for catalysis, forces in aqueous solution, carbonyl- and acyl-group reactions, practical kinetics, more. 864pp. 5⅜ x 8½.
65460-5

THE HISTORICAL BACKGROUND OF CHEMISTRY, Henry M. Leicester. Evolution of ideas, not individual biography. Concentrates on formulation of a coherent set of chemical laws. 260pp. 5⅜ x 8½. 61053-5

A SHORT HISTORY OF CHEMISTRY, J. R. Partington. Classic exposition explores origins of chemistry, alchemy, early medical chemistry, nature of atmosphere, theory of valency, laws and structure of atomic theory, much more. 428pp. 5⅜ x 8½. (Available in U.S. only.) 65977-1

GENERAL CHEMISTRY, Linus Pauling. Revised 3rd edition of classic first-year text by Nobel laureate. Atomic and molecular structure, quantum mechanics, statistical mechanics, thermodynamics correlated with descriptive chemistry. Problems. 992pp. 5⅜ x 8½. 65622-5

Engineering

DE RE METALLICA, Georgius Agricola. The famous Hoover translation of greatest treatise on technological chemistry, engineering, geology, mining of early modern times (1556). All 289 original woodcuts. 638pp. 6¾ x 11. 60006-8

FUNDAMENTALS OF ASTRODYNAMICS, Roger Bate et al. Modern approach developed by U.S. Air Force Academy. Designed as a first course. Problems, exercises. Numerous illustrations. 455pp. 5⅜ x 8½. 60061-0

DYNAMICS OF FLUIDS IN POROUS MEDIA, Jacob Bear. For advanced students of ground water hydrology, soil mechanics and physics, drainage and irrigation engineering and more. 335 illustrations. Exercises, with answers. 784pp. 6⅛ x 9¼. 65675-6

ANALYTICAL MECHANICS OF GEARS, Earle Buckingham. Indispensable reference for modern gear manufacture covers conjugate gear-tooth action, gear-tooth profiles of various gears, many other topics. 263 figures. 102 tables. 546pp. 5⅜ x 8½. 65712-4

MECHANICS, J. P. Den Hartog. A classic introductory text or refresher. Hundreds of applications and design problems illuminate fundamentals of trusses, loaded beams and cables, etc. 334 answered problems. 462pp. 5⅜ x 8½. 60754-2

MECHANICAL VIBRATIONS, J. P. Den Hartog. Classic textbook offers lucid explanations and illustrative models, applying theories of vibrations to a variety of practical industrial engineering problems. Numerous figures. 233 problems, solutions. Appendix. Index. Preface. 436pp. 5⅜ x 8½. 64785-4

STRENGTH OF MATERIALS, J. P. Den Hartog. Full, clear treatment of basic material (tension, torsion, bending, etc.) plus advanced material on engineering methods, applications. 350 answered problems. 323pp. 5⅜ x 8½. 60755-0

A HISTORY OF MECHANICS, René Dugas. Monumental study of mechanical principles from antiquity to quantum mechanics. Contributions of ancient Greeks, Galileo, Leonardo, Kepler, Lagrange, many others. 671pp. 5⅜ x 8½. 65632-2

METAL FATIGUE, N. E. Frost, K. J. Marsh, and L. P. Pook. Definitive, clearly written, and well-illustrated volume addresses all aspects of the subject, from the historical development of understanding metal fatigue to vital concepts of the cyclic stress that causes a crack to grow. Includes 7 appendixes. 544pp. 5⅜ x 8½. 40927-9

STATISTICAL MECHANICS: Principles and Applications, Terrell L. Hill. Standard text covers fundamentals of statistical mechanics, applications to fluctuation theory, imperfect gases, distribution functions, more. 448pp. 5⅜ x 8½. 65390-0

THE VARIATIONAL PRINCIPLES OF MECHANICS, Cornelius Lanczos. Graduate level coverage of calculus of variations, equations of motion, relativistic mechanics, more. First inexpensive paperbound edition of classic treatise. Index. Bibliography. 418pp. 5⅜ x 8½. 65067-7

THE VARIOUS AND INGENIOUS MACHINES OF AGOSTINO RAMELLI: A Classic Sixteenth-Century Illustrated Treatise on Technology, Agostino Ramelli. One of the most widely known and copied works on machinery in the 16th century. 194 detailed plates of water pumps, grain mills, cranes, more. 608pp. 9 x 12. 28180-9

ORDINARY DIFFERENTIAL EQUATIONS AND STABILITY THEORY: An Introduction, David A. Sánchez. Brief, modern treatment. Linear equation, stability theory for autonomous and nonautonomous systems, etc. 164pp. 5⅝ x 8¼. 63828-6

ROTARY WING AERODYNAMICS, W. Z. Stepniewski. Clear, concise text covers aerodynamic phenomena of the rotor and offers guidelines for helicopter performance evaluation. Originally prepared for NASA. 537 figures. 640pp. 6⅛ x 9¼. 64647-5

INTRODUCTION TO SPACE DYNAMICS, William Tyrrell Thomson. Comprehensive, classic introduction to space-flight engineering for advanced undergraduate and graduate students. Includes vector algebra, kinematics, transformation of coordinates. Bibliography. Index. 352pp. 5⅜ x 8½. 65113-4

HISTORY OF STRENGTH OF MATERIALS, Stephen P. Timoshenko. Excellent historical survey of the strength of materials with many references to the theories of elasticity and structure. 245 figures. 452pp. 5⅜ x 8½. 61187-6

ANALYTICAL FRACTURE MECHANICS, David J. Unger. Self-contained text supplements standard fracture mechanics texts by focusing on analytical methods for determining crack-tip stress and strain fields. 336pp. 6⅛ x 9¼. 41737-9

Mathematics

HANDBOOK OF MATHEMATICAL FUNCTIONS WITH FORMULAS, GRAPHS, AND MATHEMATICAL TABLES, edited by Milton Abramowitz and Irene A. Stegun. Vast compendium: 29 sets of tables, some to as high as 20 places. 1,046pp. 8 x 10½. 61272-4

FUNCTIONAL ANALYSIS (Second Corrected Edition), George Bachman and Lawrence Narici. Excellent treatment of subject geared toward students with background in linear algebra, advanced calculus, physics and engineering. Text covers introduction to inner-product spaces, normed, metric spaces, and topological spaces; complete orthonormal sets, the Hahn-Banach Theorem and its consequences, and many other related subjects. 1966 ed. 544pp. 6⅛ x 9¼. 40251-7

ASYMPTOTIC EXPANSIONS OF INTEGRALS, Norman Bleistein & Richard A. Handelsman. Best introduction to important field with applications in a variety of scientific disciplines. New preface. Problems. Diagrams. Tables. Bibliography. Index. 448pp. 5⅜ x 8½. 65082-0

FAMOUS PROBLEMS OF GEOMETRY AND HOW TO SOLVE THEM, Benjamin Bold. Squaring the circle, trisecting the angle, duplicating the cube: learn their history, why they are impossible to solve, then solve them yourself. 128pp. 5⅜ x 8½. 24297-8

VECTOR AND TENSOR ANALYSIS WITH APPLICATIONS, A. I. Borisenko and I. E. Tarapov. Concise introduction. Worked-out problems, solutions, exercises. 257pp. 5⅝ x 8¼. 63833-2

THE ABSOLUTE DIFFERENTIAL CALCULUS (CALCULUS OF TENSORS), Tullio Levi-Civita. Great 20th-century mathematician's classic work on material necessary for mathematical grasp of theory of relativity. 452pp. 5⅝ x 8¼. 63401-9

AN INTRODUCTION TO ORDINARY DIFFERENTIAL EQUATIONS, Earl A. Coddington. A thorough and systematic first course in elementary differential equations for undergraduates in mathematics and science, with many exercises and problems (with answers). Index. 304pp. 5⅜ x 8½. 65942-9

FOURIER SERIES AND ORTHOGONAL FUNCTIONS, Harry F. Davis. An incisive text combining theory and practical example to introduce Fourier series, orthogonal functions and applications of the Fourier method to boundary-value problems. 570 exercises. Answers and notes. 416pp. 5⅜ x 8½. 65973-9

COMPUTABILITY AND UNSOLVABILITY, Martin Davis. Classic graduate-level introduction to theory of computability, usually referred to as theory of recurrent functions. New preface and appendix. 288pp. 5⅜ x 8½. 61471-9

ASYMPTOTIC METHODS IN ANALYSIS, N. G. de Bruijn. An inexpensive, comprehensive guide to asymptotic methods–the pioneering work that teaches by explaining worked examples in detail. Index. 224pp. 5⅜ x 8½ 64221-6

ESSAYS ON THE THEORY OF NUMBERS, Richard Dedekind. Two classic essays by great German mathematician: on the theory of irrational numbers; and on transfinite numbers and properties of natural numbers. 115pp. 5⅜ x 8½. 21010-3

THE FUNCTIONS OF MATHEMATICAL PHYSICS, Harry Hochstadt. Comprehensive treatment of orthogonal polynomials, hypergeometric functions, Hill's equation, much more. Bibliography. Index. 322pp. 5⅜ x 8½. 65214-9

THREE PEARLS OF NUMBER THEORY, A. Y. Khinchin. Three compelling puzzles require proof of a basic law governing the world of numbers. Challenges concern van der Waerden's theorem, the Landau-Schnirelmann hypothesis and Mann's theorem, and a solution to Waring's problem. Solutions included. 64pp. 5⅜ x 8½. 40026-3

CALCULUS REFRESHER FOR TECHNICAL PEOPLE, A. Albert Klaf. Covers important aspects of integral and differential calculus via 756 questions. 566 problems, most answered. 431pp. 5⅜ x 8½. 20370-0

THE PHILOSOPHY OF MATHEMATICS: An Introductory Essay, Stephan Körner. Surveys the views of Plato, Aristotle, Leibniz & Kant concerning propositions and theories of applied and pure mathematics. Introduction. Two appendices. Index. 198pp. 5⅜ x 8½. 25048-2

INTRODUCTORY REAL ANALYSIS, A.N. Kolmogorov, S. V. Fomin. Translated by Richard A. Silverman. Self-contained, evenly paced introduction to real and functional analysis. Some 350 problems. 403pp. 5⅜ x 8½. 61226-0

APPLIED ANALYSIS, Cornelius Lanczos. Classic work on analysis and design of finite processes for approximating solution of analytical problems. Algebraic equations, matrices, harmonic analysis, quadrature methods, much more. 559pp. 5⅜ x 8½. 65656-X

AN INTRODUCTION TO ALGEBRAIC STRUCTURES, Joseph Landin. Superb self-contained text covers "abstract algebra": sets and numbers, theory of groups, theory of rings, much more. Numerous well-chosen examples, exercises. 247pp. 5⅜ x 8½. 65940-2

SPECIAL FUNCTIONS, N. N. Lebedev. Translated by Richard Silverman. Famous Russian work treating more important special functions, with applications to specific problems of physics and engineering. 38 figures. 308pp. 5⅜ x 8½. 60624-4

QUALITATIVE THEORY OF DIFFERENTIAL EQUATIONS, V. V. Nemytskii and V.V. Stepanov. Classic graduate-level text by two prominent Soviet mathematicians covers classical differential equations as well as topological dynamics and ergodic theory. Bibliographies. 523pp. 5⅜ x 8½. 65954-2

NUMBER THEORY AND ITS HISTORY, Oystein Ore. Unusually clear, accessible introduction covers counting, properties of numbers, prime numbers, much more. Bibliography. 380pp. 5⅜ x 8½. 65620-9

THEORY OF MATRICES, Sam Perlis. Outstanding text covering rank, nonsingularity and inverses in connection with the development of canonical matrices under the relation of equivalence, and without the intervention of determinants. Includes exercises. 237pp. 5⅜ x 8½. 66810-X

INTRODUCTION TO ANALYSIS, Maxwell Rosenlicht. Unusually clear, accessible coverage of set theory, real number system, metric spaces, continuous functions, Riemann integration, multiple integrals, more. Wide range of problems. Undergraduate level. Bibliography. 254pp. 5⅜ x 8½. 65038-3

MODERN NONLINEAR EQUATIONS, Thomas L. Saaty. Emphasizes practical solution of problems; covers seven types of equations. ". . . a welcome contribution to the existing literature...."—*Math Reviews.* 490pp. 5⅜ x 8½. 64232-1

MATRICES AND LINEAR ALGEBRA, Hans Schneider and George Phillip Barker. Basic textbook covers theory of matrices and its applications to systems of linear equations and related topics such as determinants, eigenvalues and differential equations. Numerous exercises. 432pp. 5⅜ x 8½. 66014-1

MATHEMATICS APPLIED TO CONTINUUM MECHANICS, Lee A. Segel. Analyzes models of fluid flow and solid deformation. For upper-level math, science and engineering students. 608pp. 5⅜ x 8½. 65369-2

ELEMENTS OF REAL ANALYSIS, David A. Sprecher. Classic text covers fundamental concepts, real number system, point sets, functions of a real variable, Fourier series, much more. Over 500 exercises. 352pp. 5⅜ x 8½. 65385-4

AN INTRODUCTION TO MATRICES, SETS AND GROUPS FOR SCIENCE STUDENTS, G. Stephenson. Concise, readable text introduces sets, groups, and most importantly, matrices to undergraduate students of physics, chemistry, and engineering. Problems. 164pp. 5⅜ x 8½. 65077-4

SET THEORY AND LOGIC, Robert R. Stoll. Lucid introduction to unified theory of mathematical concepts. Set theory and logic seen as tools for conceptual understanding of real number system. 496pp. 5⅜ x 8¼. 63829-4

TENSOR CALCULUS, J.L. Synge and A. Schild. Widely used introductory text covers spaces and tensors, basic operations in Riemannian space, non-Riemannian spaces, etc. 324pp. 5⅜ x 8¼. 63612-7

ORDINARY DIFFERENTIAL EQUATIONS, Morris Tenenbaum and Harry Pollard. Exhaustive survey of ordinary differential equations for undergraduates in mathematics, engineering, science. Thorough analysis of theorems. Diagrams. Bibliography. Index. 818pp. 5⅜ x 8½. 64940-7

INTEGRAL EQUATIONS, F. G. Tricomi. Authoritative, well-written treatment of extremely useful mathematical tool with wide applications. Volterra Equations, Fredholm Equations, much more. Advanced undergraduate to graduate level. Exercises. Bibliography. 238pp. 5⅜ x 8½. 64828-1

FOURIER SERIES, Georgi P. Tolstov. Translated by Richard A. Silverman. A valuable addition to the literature on the subject, moving clearly from subject to subject and theorem to theorem. 107 problems, answers. 336pp. 5⅜ x 8½. 63317-9

POPULAR LECTURES ON MATHEMATICAL LOGIC, Hao Wang. Noted logician's lucid treatment of historical developments, set theory, model theory, recursion theory and constructivism, proof theory, more. 3 appendixes. Bibliography. 1981 edition. ix + 283pp. 5⅜ x 8½. 67632-3

CALCULUS OF VARIATIONS, Robert Weinstock. Basic introduction covering isoperimetric problems, theory of elasticity, quantum mechanics, electrostatics, etc. Exercises throughout. 326pp. 5⅜ x 8½. 63069-2

THE CONTINUUM: A Critical Examination of the Foundation of Analysis, Hermann Weyl. Classic of 20th-century foundational research deals with the conceptual problem posed by the continuum. 156pp. 5⅜ x 8½. 67982-9

CHALLENGING MATHEMATICAL PROBLEMS WITH ELEMENTARY SOLUTIONS, A. M. Yaglom and I. M. Yaglom. Over 170 challenging problems on probability theory, combinatorial analysis, points and lines, topology, convex polygons, many other topics. Solutions. Total of 445pp. 5⅜ x 8½. Two-vol. set.
Vol. I: 65536-9 Vol. II: 65537-7

A SURVEY OF NUMERICAL MATHEMATICS, David M. Young and Robert Todd Gregory. Broad self-contained coverage of computer-oriented numerical algorithms for solving various types of mathematical problems in linear algebra, ordinary and partial, differential equations, much more. Exercises. Total of 1,248pp. 5⅜ x 8½. Two volumes.
Vol. I: 65691-8 Vol. II: 65692-6

INTRODUCTION TO PARTIAL DIFFERENTIAL EQUATIONS WITH APPLICATIONS, E. C. Zachmanoglou and Dale W. Thoe. Essentials of partial differential equations applied to common problems in engineering and the physical sciences. Problems and answers. 416pp. 5⅜ x 8½. 65251-3

THE THEORY OF GROUPS, Hans J. Zassenhaus. Well-written graduate-level text acquaints reader with group-theoretic methods and demonstrates their usefulness in mathematics. Axioms, the calculus of complexes, homomorphic mapping, p-group theory, more. Many proofs shorter and more transparent than older ones. 276pp. 5⅜ x 8½. 40922-8

DISTRIBUTION THEORY AND TRANSFORM ANALYSIS: An Introduction to Generalized Functions, with Applications, A. H. Zemanian. Provides basics of distribution theory, describes generalized Fourier and Laplace transformations. Numerous problems. 384pp. 5⅜ x 8½. 65479-6

Math–Decision Theory, Statistics, Probability

ELEMENTARY DECISION THEORY, Herman Chernoff and Lincoln E. Moses. Clear introduction to statistics and statistical theory covers data processing, probability and random variables, testing hypotheses, much more. Exercises. 364pp. 5⅜ x 8½. 65218-1

STATISTICS MANUAL, Edwin L. Crow et al. Comprehensive, practical collection of classical and modern methods prepared by U.S. Naval Ordnance Test Station. Stress on use. Basics of statistics assumed. 288pp. 5⅜ x 8½. 60599-X

SOME THEORY OF SAMPLING, William Edwards Deming. Analysis of the problems, theory and design of sampling techniques for social scientists, industrial managers and others who find statistics important at work. 61 tables. 90 figures. xvii +602pp. 5⅜ x 8½. 64684-X

STATISTICAL ADJUSTMENT OF DATA, W. Edwards Deming. Introduction to basic concepts of statistics, curve fitting, least squares solution, conditions without parameter, conditions containing parameters. 26 exercises worked out. 271pp. 5⅜ x 8½. 64685-8

LINEAR PROGRAMMING AND ECONOMIC ANALYSIS, Robert Dorfman, Paul A. Samuelson and Robert M. Solow. First comprehensive treatment of linear programming in standard economic analysis. Game theory, modern welfare economics, Leontief input-output, more. 525pp. 5⅜ x 8½. 65491-5

DICTIONARY/OUTLINE OF BASIC STATISTICS, John E. Freund and Frank J. Williams. A clear concise dictionary of over 1,000 statistical terms and an outline of statistical formulas covering probability, nonparametric tests, much more. 208pp. 5⅜ x 8½. 66796-0

PROBABILITY: An Introduction, Samuel Goldberg. Excellent basic text covers set theory, probability theory for finite sample spaces, binomial theorem, much more. 360 problems. Bibliographies. 322pp. 5⅜ x 8½. 65252-1

GAMES AND DECISIONS: Introduction and Critical Survey, R. Duncan Luce and Howard Raiffa. Superb nontechnical introduction to game theory, primarily applied to social sciences. Utility theory, zero-sum games, n-person games, decision-making, much more. Bibliography. 509pp. 5⅜ x 8½. 65943-7

FIFTY CHALLENGING PROBLEMS IN PROBABILITY WITH SOLUTIONS, Frederick Mosteller. Remarkable puzzlers, graded in difficulty, illustrate elementary and advanced aspects of probability. Detailed solutions. 88pp. 5⅜ x 8½. 65355-2

PROBABILITY THEORY: A Concise Course, Y. A. Rozanov. Highly readable, self-contained introduction covers combination of events, dependent events, Bernoulli trials, etc. 148pp. 5⅜ x 8¼. 63544-9

STATISTICAL METHOD FROM THE VIEWPOINT OF QUALITY CONTROL, Walter A. Shewhart. Important text explains regulation of variables, uses of statistical control to achieve quality control in industry, agriculture, other areas. 192pp. 5⅜ x 8½. 65232-7

THE COMPLEAT STRATEGYST: Being a Primer on the Theory of Games of Strategy, J. D. Williams. Highly entertaining classic describes, with many illustrated examples, how to select best strategies in conflict situations. Prefaces. Appendices. 268pp. 5⅜ x 8½. 25101-2

Math–Geometry and Topology

ELEMENTARY CONCEPTS OF TOPOLOGY, Paul Alexandroff. Elegant, intuitive approach to topology from set-theoretic topology to Betti groups; how concepts of topology are useful in math and physics. 25 figures. 57pp. 5⅜ x 8½. 60747-X

COMBINATORIAL TOPOLOGY, P. S. Alexandrov. Clearly written, well-organized, three-part text begins by dealing with certain classic problems without using the formal techniques of homology theory and advances to the central concept, the Betti groups. Numerous detailed examples. 654pp. 5⅜ x 8½. 40179-0

EXPERIMENTS IN TOPOLOGY, Stephen Barr. Classic, lively explanation of one of the byways of mathematics. Klein bottles, Moebius strips, projective planes, map coloring, problem of the Koenigsberg bridges, much more, described with clarity and wit. 43 figures. 210pp. 5⅜ x 8½. 25933-1

CONFORMAL MAPPING ON RIEMANN SURFACES, Harvey Cohn. Lucid, insightful book presents ideal coverage of subject. 334 exercises make book perfect for self-study. 55 figures. 352pp. 5⅝ x 8¼. 64025-6

THE GEOMETRY OF RENÉ DESCARTES, René Descartes. The great work founded analytical geometry. Original French text, Descartes's own diagrams, together with definitive Smith-Latham translation. 244pp. 5⅜ x 8½. 60068-8

THE THIRTEEN BOOKS OF EUCLID'S ELEMENTS, translated with introduction and commentary by Sir Thomas L. Heath. Definitive edition. Textual and linguistic notes, mathematical analysis. 2,500 years of critical commentary. Unabridged. 1,414pp. 5⅜ x 8½. Three-vol. set.
Vol. I: 60088-2 Vol. II: 60089-0 Vol. III: 60090-4

GEOMETRY OF COMPLEX NUMBERS, Hans Schwerdtfeger. Illuminating, widely praised book on analytic geometry of circles, the Moebius transformation, and two-dimensional non-Euclidean geometries. 200pp. 5⅝ x 8¼. 63830-8

DIFFERENTIAL GEOMETRY, Heinrich W. Guggenheimer. Local differential geometry as an application of advanced calculus and linear algebra. Curvature, transformation groups, surfaces, more. Exercises. 62 figures. 378pp. 5⅜ x 8½. 63433-7

CURVATURE AND HOMOLOGY: Enlarged Edition, Samuel I. Goldberg. Revised edition examines topology of differentiable manifolds; curvature, homology of Riemannian manifolds; compact Lie groups; complex manifolds; curvature, homology of Kaehler manifolds. New Preface. Four new appendixes. 416pp. 5⅜ x 8½. 40207-X

TOPOLOGY, John G. Hocking and Gail S. Young. Superb one-year course in classical topology. Topological spaces and functions, point-set topology, much more. Examples and problems. Bibliography. Index. 384pp. 5⅝ x 8¼. 65676-4

LECTURES ON CLASSICAL DIFFERENTIAL GEOMETRY, Second Edition, Dirk J. Struik. Excellent brief introduction covers curves, theory of surfaces, fundamental equations, geometry on a surface, conformal mapping, other topics. Problems. 240pp. 5⅜ x 8½. 65609-8

Math–History of

A SHORT ACCOUNT OF THE HISTORY OF MATHEMATICS, W. W. Rouse Ball. One of clearest, most authoritative surveys from the Egyptians and Phoenicians through 19th-century figures such as Grassman, Galois, Riemann. Fourth edition. 522pp. 5⅜ x 8½. 20630-0

THE HISTORY OF THE CALCULUS AND ITS CONCEPTUAL DEVELOPMENT, Carl B. Boyer. Origins in antiquity, medieval contributions, work of Newton, Leibniz, rigorous formulation. Treatment is verbal. 346pp. 5⅜ x 8½. 60509-4

THE HISTORICAL ROOTS OF ELEMENTARY MATHEMATICS, Lucas N. H. Bunt, Phillip S. Jones, and Jack D. Bedient. Fundamental underpinnings of modern arithmetic, algebra, geometry and number systems derived from ancient civilizations. 320pp. 5⅜ x 8½. 25563-8

A HISTORY OF MATHEMATICAL NOTATIONS, Florian Cajori. This classic study notes the first appearance of a mathematical symbol and its origin, the competition it encountered, its spread among writers in different countries, its rise to popularity, its eventual decline or ultimate survival. Original 1929 two-volume edition presented here in one volume. xxviii+820pp. 5⅜ x 8½. 67766-4

GAMES, GODS & GAMBLING: A History of Probability and Statistical Ideas, F. N. David. Episodes from the lives of Galileo, Fermat, Pascal, and others illustrate this fascinating account of the roots of mathematics. Features thought-provoking references to classics, archaeology, biography, poetry. 1962 edition. 304pp. 5⅜ x 8½. (Available in U.S. only.) 40023-9

OF MEN AND NUMBERS: The Story of the Great Mathematicians, Jane Muir. Fascinating accounts of the lives and accomplishments of history's greatest mathematical minds–Pythagoras, Descartes, Euler, Pascal, Cantor, many more. Anecdotal, illuminating. 30 diagrams. Bibliography. 256pp. 5⅜ x 8½. 28973-7

HISTORY OF MATHEMATICS, David E. Smith. Nontechnical survey from ancient Greece and Orient to late 19th century; evolution of arithmetic, geometry, trigonometry, calculating devices, algebra, the calculus. 362 illustrations. 1,355pp. 5⅜ x 8½. Two-vol. set. Vol. I: 20429-4 Vol. II: 20430-8

A CONCISE HISTORY OF MATHEMATICS, Dirk J. Struik. The best brief history of mathematics. Stresses origins and covers every major figure from ancient Near East to 19th century. 41 illustrations. 195pp. 5⅜ x 8½. 60255-9

Physics

OPTICAL RESONANCE AND TWO-LEVEL ATOMS, L. Allen and J. H. Eberly. Clear, comprehensive introduction to basic principles behind all quantum optical resonance phenomena. 53 illustrations. Preface. Index. 256pp. 5⅜ x 8½. 65533-4

ULTRASONIC ABSORPTION: An Introduction to the Theory of Sound Absorption and Dispersion in Gases, Liquids and Solids, A. B. Bhatia. Standard reference in the field provides a clear, systematically organized introductory review of fundamental concepts for advanced graduate students, research workers. Numerous diagrams. Bibliography. 440pp. 5⅜ x 8½. 64917-2

QUANTUM THEORY, David Bohm. This advanced undergraduate-level text presents the quantum theory in terms of qualitative and imaginative concepts, followed by specific applications worked out in mathematical detail. Preface. Index. 655pp. 5⅜ x 8½. 65969-0

ATOMIC PHYSICS (8th edition), Max Born. Nobel laureate's lucid treatment of kinetic theory of gases, elementary particles, nuclear atom, wave-corpuscles, atomic structure and spectral lines, much more. Over 40 appendices, bibliography. 495pp. 5⅜ x 8½. 65984-4

AN INTRODUCTION TO HAMILTONIAN OPTICS, H. A. Buchdahl. Detailed account of the Hamiltonian treatment of aberration theory in geometrical optics. Many classes of optical systems defined in terms of the symmetries they possess. Problems with detailed solutions. 1970 edition. xv + 360pp. 5⅜ x 8½. 67597-1

THIRTY YEARS THAT SHOOK PHYSICS: The Story of Quantum Theory, George Gamow. Lucid, accessible introduction to influential theory of energy and matter. Careful explanations of Dirac's anti-particles, Bohr's model of the atom, much more. 12 plates. Numerous drawings. 240pp. 5⅜ x 8½. 24895-X

ELECTRONIC STRUCTURE AND THE PROPERTIES OF SOLIDS: The Physics of the Chemical Bond, Walter A. Harrison. Innovative text offers basic understanding of the electronic structure of covalent and ionic solids, simple metals, transition metals and their compounds. Problems. 1980 edition. 582pp. 6⅛ x 9¼.
66021-4

HYDRODYNAMIC AND HYDROMAGNETIC STABILITY, S. Chandrasekhar. Lucid examination of the Rayleigh-Benard problem; clear coverage of the theory of instabilities causing convection. 704pp. 5⅜ x 8¼. 64071-X

INVESTIGATIONS ON THE THEORY OF THE BROWNIAN MOVEMENT, Albert Einstein. Five papers (1905–8) investigating dynamics of Brownian motion and evolving elementary theory. Notes by R. Fürth. 122pp. 5⅜ x 8½. 60304-0

THE PHYSICS OF WAVES, William C. Elmore and Mark A. Heald. Unique overview of classical wave theory. Acoustics, optics, electromagnetic radiation, more. Ideal as classroom text or for self-study. Problems. 477pp. 5⅜ x 8½. 64926-1

CATALOG OF DOVER BOOKS

PHYSICAL PRINCIPLES OF THE QUANTUM THEORY, Werner Heisenberg. Nobel Laureate discusses quantum theory, uncertainty, wave mechanics, work of Dirac, Schroedinger, Compton, Wilson, Einstein, etc. 184pp. 5⅜ x 8½.   60113-7

ATOMIC SPECTRA AND ATOMIC STRUCTURE, Gerhard Herzberg. One of best introductions; especially for specialist in other fields. Treatment is physical rather than mathematical. 80 illustrations. 257pp. 5⅜ x 8½.   60115-3

AN INTRODUCTION TO STATISTICAL THERMODYNAMICS, Terrell L. Hill. Excellent basic text offers wide-ranging coverage of quantum statistical mechanics, systems of interacting molecules, quantum statistics, more. 523pp. 5⅜ x 8½.   65242-4

THEORETICAL PHYSICS, Georg Joos, with Ira M. Freeman. Classic overview covers essential math, mechanics, electromagnetic theory, thermodynamics, quantum mechanics, nuclear physics, other topics. First paperback edition. xxiii + 885pp. 5⅜ x 8½.   65227-0

PROBLEMS AND SOLUTIONS IN QUANTUM CHEMISTRY AND PHYSICS, Charles S. Johnson, Jr. and Lee G. Pedersen. Unusually varied problems, detailed solutions in coverage of quantum mechanics, wave mechanics, angular momentum, molecular spectroscopy, more. 280 problems plus 139 supplementary exercises. 430pp. 6½ x 9¼.   65236-X

THEORETICAL SOLID STATE PHYSICS, Vol. 1: Perfect Lattices in Equilibrium; Vol. II: Non-Equilibrium and Disorder, William Jones and Norman H. March. Monumental reference work covers fundamental theory of equilibrium properties of perfect crystalline solids, non-equilibrium properties, defects and disordered systems. Appendices. Problems. Preface. Diagrams. Index. Bibliography. Total of 1,301pp. 5⅜ x 8½. Two volumes.   Vol. I: 65015-4   Vol. II: 65016-2

A TREATISE ON ELECTRICITY AND MAGNETISM, James Clerk Maxwell. Important foundation work of modern physics. Brings to final form Maxwell's theory of electromagnetism and rigorously derives his general equations of field theory. 1,084pp. 5⅜ x 8½. Two-vol. set.   Vol. I: 60636-8   Vol. II: 60637-6

OPTICKS, Sir Isaac Newton. Newton's own experiments with spectroscopy, colors, lenses, reflection, refraction, etc., in language the layman can follow. Foreword by Albert Einstein. 532pp. 5⅜ x 8½.   60205-2

THEORY OF ELECTROMAGNETIC WAVE PROPAGATION, Charles Herach Papas. Graduate-level study discusses the Maxwell field equations, radiation from wire antennas, the Doppler effect and more. xiii + 244pp. 5⅜ x 8½.   65678-5

INTRODUCTION TO QUANTUM MECHANICS With Applications to Chemistry, Linus Pauling & E. Bright Wilson, Jr. Classic undergraduate text by Nobel Prize winner applies quantum mechanics to chemical and physical problems. Numerous tables and figures enhance the text. Chapter bibliographies. Appendices. Index. 468pp. 5⅜ x 8½.   64871-0